BOTANICAL RESEARCH AND PRACTICES

SPRUCE

ECOLOGY, MANAGEMENT AND CONSERVATION

BOTANICAL RESEARCH AND PRACTICES

Additional books in this series can be found on Nova's website under the Series tab.

Additional E-books in this series can be found on Nova's website under the E-book tab.

ENVIRONMENTAL SCIENCE, ENGINEERING AND TECHNOLOGY

Additional books in this series can be found on Nova's website under the Series tab.

Additional E-books in this series can be found on Nova's website under the E-book tab.

BOTANICAL RESEARCH AND PRACTICES

SPRUCE

ECOLOGY, MANAGEMENT AND CONSERVATION

KAJETAN I. NOWAK
AND
HELENA F. STRYBEL
EDITORS

Nova Science Publishers, Inc.
New York

Copyright © 2012 by Nova Science Publishers, Inc.

All rights reserved. No part of this book may be reproduced, stored in a retrieval system or transmitted in any form or by any means: electronic, electrostatic, magnetic, tape, mechanical photocopying, recording or otherwise without the written permission of the Publisher.

For permission to use material from this book please contact us:
Telephone 631-231-7269; Fax 631-231-8175
Web Site: http://www.novapublishers.com

NOTICE TO THE READER

The Publisher has taken reasonable care in the preparation of this book, but makes no expressed or implied warranty of any kind and assumes no responsibility for any errors or omissions. No liability is assumed for incidental or consequential damages in connection with or arising out of information contained in this book. The Publisher shall not be liable for any special, consequential, or exemplary damages resulting, in whole or in part, from the readers' use of, or reliance upon, this material. Any parts of this book based on government reports are so indicated and copyright is claimed for those parts to the extent applicable to compilations of such works.

Independent verification should be sought for any data, advice or recommendations contained in this book. In addition, no responsibility is assumed by the publisher for any injury and/or damage to persons or property arising from any methods, products, instructions, ideas or otherwise contained in this publication.

This publication is designed to provide accurate and authoritative information with regard to the subject matter covered herein. It is sold with the clear understanding that the Publisher is not engaged in rendering legal or any other professional services. If legal or any other expert assistance is required, the services of a competent person should be sought. FROM A DECLARATION OF PARTICIPANTS JOINTLY ADOPTED BY A COMMITTEE OF THE AMERICAN BAR ASSOCIATION AND A COMMITTEE OF PUBLISHERS.

Additional color graphics may be available in the e-book version of this book.

Library of Congress Cataloging-in-Publication Data
Spruce : ecology, management, and conservation / editors, Kajetan I. Nowak and Helena F. Strybel.
   p. cm.
 Includes bibliographical references and index.
 ISBN 978-1-61942-494-4 (soft cover)
  1. Spruce--Ecology. 2. Spruce--Conservation. 3. Forest ecology. 4. Forest conservation. 5. Forest management. I. Nowak, Kajetan I. II. Strybel, Helena F.
 QK494.5.P66S67 2011
 585'.2--dc23
                    2011048506

Published by Nova Science Publishers, Inc. † New York

Contents

Preface		**vii**
Chapter 1	Norway Spruce *Picea Abies* Regeneration and Canopy Disturbance in a Carpathian Subalpine Forest *Jan Holeksa, Tomasz Zielonka and Magdalena Żywiec*	**1**
Chapter 2	Uncultured Archaea in Spruce Rhizospheres and Mycorrhizas *Malin Bomberg*	**41**
Chapter 3	Reversible Variations in Some Wood Properties of Norway Spruce (*Picea Abies* Karst.), Depending on the Tree Felling Date *Ernst Zürcher, Christian Rogenmoser, Azadeh Soleimany Kartalaei, and Diane Rambert*	**75**
Chapter 4	Exploration of Forest Vegetation with Dynamic Modeling, GIS and Remote Sensing *Lubos Matejicek*	**95**
Chapter 5	Vitamin C as a Stress Bioindicator of Norway Spruce: A Case Study in Slovenia *Samar Al Sayegh Petkovšek and Boštjan Pokorny*	**115**
Index		**135**

PREFACE

In this book, the authors present current research in the study of the ecology, management and conservation of spruce. Topics discussed include Norway spruce picea abies regeneration and canopy disturbance in a carpathian subalpine forest; uncultured archaea in spruce rhizospheres and mycorrhizas; reversible variations in some wood properties of Norway spruce depending on the tree felling date; Vitamin C as a stress bioindicator of Norway spruce and landscape spruce planning management.

Chapter 1 – Subject. In this chapter, the authors focus on the relationship between the recruitment of Norway spruce *Picea abies* seedlings and saplings and canopy disturbances. Special attention was paid to canopy gaps, soil disturbances caused by windthrows, deadwood deposited on the forest floor, and the heterogeneity of field-layer vegetation as factors conditioning spruce regeneration. They were particularly interested in regeneration delay after stand disturbances, which seems to be of importance for dynamics of subalpine spruce forests.

Location. The study was carried out in the subalpine spruce forest on the Babia Góra Mt. located in the Western Carpathian Mountains. The forest examined has been under protection since 1930 and no management and timber extraction have been carried out since that time. Data was collected in 14,4 ha study plot within altitudinal zone 1188 – 1300 m a.s.l.

Results. The recruitment of young spruces is related to the size of canopy gaps being more abundant in large gaps than in small ones. Deadwood and windthrow mounds are the most important sites for spruce regeneration. They support as little as 1.5% of the germinants, but their contribution is 40 times higher for saplings. However, they do not offer favorable conditions for regeneration of spruce immediately after tree death. Deadwood and windthrow mounds are available for seedlings not earlier than about 20 years after tree death. Their abundant recruitment lasts for the next 20-40 years. Thick logs are essential for establishment of spruces. The recruitment of seedlings on soil depends on the field-layer dominants. It is most successful in patches of *Vaccinium myrtillus*, but in gaps it is effectively inhibited by fern, *Athyrium distentifolium*. The annual height growth rate of spruces growing on deadwood in large gaps > 400 m2 ranged between 2.2 cm for individuals 30 cm tall and 15.5 cm for 200-cm individuals. Spruces of 30 cm growing on CWD in large gaps reach the height of 5 m after about 40 years.

Conclusions. Gaps cannot solely promote spruce regeneration, because of the unfavorable changes initiated in the field-layer vegetation after canopy opening. Disturbed soil and coarse woody debris deposited in gaps are indispensable for spruce recruitment even in gaps. Without windthrow mounds and large-sized downed wood, regeneration would rather be reduced in large gaps after their creation, because of exuberant growth of *A. distentifolium* and other species. Slow decomposition of coarse woody debris and a low rate of height growth of young spruces result in the long temporal gap of about 50 years between disturbance and establishment of new generation of spruce saplings. Because of the temporal gap between the disturbance and the seedling recruitment and a slow growth rate of young spruces, a lag of about 100 years between disturbance and establishment of 5-m tall young stand can be expected in the Carpathian subalpine forests.

Chapter 2 – Archaea are microorganisms belonging to the third domain of cellular life on earth. Despite being unicellular microorganisms, they differ from the bacteria in many ways, and actually share an evolutionary history with the Eukaryotic branch of life. Since their discovery in extreme habitats, they have during the last two decades been found to inhabit almost all environments on Earth. Specific linages of archaea have been found to live forest soil ecosystems, and they have been found to be especially associated with boreal forest tree roots and mycorrhizospheres. These archaea belong to the so called Group I.1c of uncultured Thaumarchaeota (formerly included in the Phylum Crenarchaeota), and euryarchaeotal linages phylogenetically falling with the generally extremely halophilic Halobacteriales, as well as with methylotrophic and acetoclastic methanogens belonging to the Methanolobus and Methanosaeta, respectively.

Norway spruce is one of the most common forest trees in the Fennoscandian boreal forest. Only few studies have thus far been conducted on archaea inhabiting the rhizosphere and mycorrhizosphere of boreal forest trees, and even fewer have concentrated on Norway spruce. However, it has been shown that the tree species has profound effect on the community composition of the archaea in the rhizosphere. When the tree roots are colonized by ectomycorrhizal fungi, the effect of the fungus dominates and the effect of the tree decreases. A spruce seedling grown in natural humus from a Scots pine stand gathered more detectable archaea in its roots than a Scots pine seedling, but when grown in natural humus from a Norway spruce stand the detectable number of archaea in the Norway spruce roots was much reduced. It has been shown that different tree species affect the soil they live in, and that Norway spruce has a tendency to acidify the soil, while for example silver birch increase the soil pH. Acidic spruce needle litter also bring recalcitrant organic matter and phenolics to the soil, while birch litter increase soil organic matter and nitrogen. An even greater difference between tree species can be detected in the composition of bacterial groups in the rhizospheres and mycorrhizospheres of different boreal forest tree species. The fact that archaea are generally found only in the roots and rhizosphere, and not in the soil uncolonized by mycorrhizal fungi or tree roots indicate a relationship between the archaea and the ectomycorrhizal fungi and the tree.

This chapter will concentrate on the archaea detected in the roots and mycorrhizosphere of boreal forest Norway spruce in comparison to other boreal tree species, but will also shortly touch on the subject of bacteria.

Chapter 3 – Traditional knowledge, in the form of so-called rural rules, indicates that the date of tree felling has an important influence on wood quality. The main factor, after the season of the year, is said to be the position of the moon. The object of the research presented here was to study the variability of some user-related properties of wood, by analyzing measurable parameters. The material stems from four different Swiss sites and is representative of central European conditions. The study involved 576 trees — Norway Spruce (*Picea abies* Karst.) and Sweet Chestnut (*Castanea sativa* Mill.) — felled on 48 dates throughout the fall and spring of 2003–2004 (always on Mondays or Thursdays). Before the start of the experiment, one sample was taken on the same day from each of the tested trees, to serve as reference. Wood properties analyzed are: water-loss, shrinkage under controlled drying, air dry and oven dry density. The statistical analysis of the complete data series reveals (in addition to a seasonal trend) a generally weak, but highly significant role of the synodic and sidereal moon cycles and, to a lesser extent, the tropical cycle.

The lunar-related differences are more marked for the middle months of the trial. The most obvious variation in Spruce occurs between samples of trees felled immediately before the Full Moon and the samples immediately following Full Moon. Smaller series of Spruce samples were tested on hygroscopicity, compression strength and calorimetry. Here too, the strong value shifts observed around the Full Moon found a clear confirmation. The main variation factor for water uptake is however the type of forest and the site, a naturally grown mountain forest producing a clearly less hygroscopic wood. The results from this study bring some transparency and objectivity into a mainly unexplored field of traditional knowledge, a field subject to controversial discussions. Further research in chronobiology of wood could lead to an ecological technique enhancing specific wood properties.

Chapter 4 – Landscape spruce planning operates on the interface between management actions and ecosystems. It requires understanding the existing ecological state of the spruce forest and projecting vegetation over time and space to accomplish management objectives. New processing tools are needed to provide more complex spatio-temporal analysis. Nowadays, Geographic Information Systems (GISs) extended by remote sensing techniques, and terrain measurements with Global Positioning Systems (GPS) offer advanced methods for landscape spruce planning and management. Using satellite images and aerial photographs is presented as a revolutionary approach that is expanded by a next generation of sensors with the improved spectral and spatial resolution. Remote sensing has been a valuable source of information for a few decades in mapping and monitoring forest activities. Exploratory spatial data analysis and dynamic modeling enable to evaluate spatial relationships in soil, water, and wildlife resources. In order to demonstrate advanced processing tools focused on spruce ecology, management and conservation, a case study dealing with the spatio-temporal modeling of natural regeneration is carried out in the GIS that can provide a high-quality spatial database. It describes the state of the experimental areas of interest in the spruce forest, and includes a spatio-temporal model to create alternatives for planning and management. The numerical model is based on a large set of ordinary differential equations that can solve dynamic processes and spatial relationships in selected microsites. The simulation results can show the short-time succession for a regeneration decade and approximate long-term development. The use of GIS offers visualization of model outputs that offer to present the decision-making processes in a more illustrative way.

Chapter 5 – Physiological condition of Norway spruce (Picea abies (L.) Karst.) and consequently vitality of forest ecosystems was intensively studied in the period 1991 – 2007 in the northern Slovenia, i.e. in area, influenced by the Šoštanj Thermal Power Plant (ŠTPP). ŠTPP, which is the largest Slovene thermal power plant, used to be the largest Slovene emission source of gaseous pollutants (e.g. SO2, NOx), and very important source of different inorganic (e.g. heavy metals) as well as organic toxic substances (e.g. PAHs). However, extremely high SO2 emission (up to 86,000 t in 1993, and > 120.000 in 1980's, respectively) and dust emissions (up to 8,000 in 1993), have been dramatically reduced after the installation of desulphurization devices in late 1990's. Indeed in the comparison with 1993, SO2 emissions in 2007 were reduced for more then 15-folds and dust emissions for more then 35-folds, respectively. These extreme exposures in the past as well as huge changes in environmental pollution during last two decades have significantly influenced vitality of forest ecosystems including physiological conditions (e.g. contents of antioxidant) of different tree species in the study area. Therefore, vitamin C (ascorbic acid) as a sensitive, non-specific bioindicator of stress caused either by anthropogenic (e.g. air pollution) or natural stressors (climatic conditions, diseases, altitude gradient, etc) was included in a permanent survey of forest condition in northern Slovenia. Atmospheric pollutants such as ozone and sulphur dioxide cause formation of free radicals, which are involved in oxidation of proteins and lipids and injury of plant tissues. Plant cells have evolved a special detoxification defence system to cope with radicals, including formation of water-soluble antioxidant, such as vitamin C. The most significant findings and conclusions of the present study are as follows: (a) Vitamin C is a good bioindicator of oxidative stress and an early-warning tool to detect changes in the metabolism of spruce needles, although the authors found untypical reaction of antioxidant defence in the case of extremely high SO2 exposure. (b) Metabolic processes in spruce needles react to air pollution according to severity of pollution and the time of exposure. However, if spruce trees were exposed to high SO2 ambient levels and/or for a long period of time, the antioxidant defence mechanism would be damaged and the content of vitamin C would not increase as expected. (c) Lower exposure to ambient pollution results in better vitality of trees (e.g. higher contents of total (a + b) chlorophyll), as well as in rising of their defence capabilities (higher contents of vitamin C). (d) Physiological condition of Norway spruce in northern Slovenia has significantly improved since 1995, when the desulphurization devices were built on the ŠTPP, and when emissions of SO2 as well as heavy metals started dramatically and continuously decreasing in this part of Slovenia.

In: Spruce
Editors: K. I. Nowak and H. F. Strybel © 2012 Nova Science Publishers, Inc.

ISBN 978-1-61942-494-4

Chapter 1

NORWAY SPRUCE *PICEA ABIES* REGENERATION AND CANOPY DISTURBANCE IN A CARPATHIAN SUBALPINE FOREST

Jan Holeksa[1], Tomasz Zielonka and Magdalena Żywiec

W. Szafer Institute of Botany,
Polish Academy of Sciences, Kraków, Poland

ABSTRACT

Subject. In this paper we focused on the relationship between the recruitment of Norway spruce *Picea abies* seedlings and saplings and canopy disturbances. Special attention was paid to canopy gaps, soil disturbances caused by windthrows, deadwood deposited on the forest floor, and the heterogeneity of field-layer vegetation as factors conditioning spruce regeneration. We were particularly interested in regeneration delay after stand disturbances, which seems to be of importance for dynamics of subalpine spruce forests.

Location. The study was carried out in the subalpine spruce forest on the Babia Góra Mt. located in the Western Carpathian Mountains. The forest examined has been under protection since 1930 and no

---

[1] e-mail: j.holeksa@botany.pl

management and timber extraction have been carried out since that time. Data was collected in 14,4 ha study plot within altitudinal zone 1188 – 1300 m a.s.l.

Results. The recruitment of young spruces is related to the size of canopy gaps being more abundant in large gaps than in small ones. Deadwood and windthrow mounds are the most important sites for spruce regeneration. They support as little as 1.5% of the germinants, but their contribution is 40 times higher for saplings. However, they do not offer favorable conditions for regeneration of spruce immediately after tree death. Deadwood and windthrow mounds are available for seedlings not earlier than about 20 years after tree death. Their abundant recruitment lasts for the next 20-40 years. Thick logs are essential for establishment of spruces. The recruitment of seedlings on soil depends on the field-layer dominants. It is most successful in patches of *Vaccinium myrtillus*, but in gaps it is effectively inhibited by fern, *Athyrium distentifolium*. The annual height growth rate of spruces growing on deadwood in large gaps > 400 m2 ranged between 2.2 cm for individuals 30 cm tall and 15.5 cm for 200-cm individuals. Spruces of 30 cm growing on CWD in large gaps reach the height of 5 m after about 40 years.

Conclusions. Gaps cannot solely promote spruce regeneration, because of the unfavorable changes initiated in the field-layer vegetation after canopy opening. Disturbed soil and coarse woody debris deposited in gaps are indispensable for spruce recruitment even in gaps. Without windthrow mounds and large-sized downed wood, regeneration would rather be reduced in large gaps after their creation, because of exuberant growth of *A. distentifolium* and other species. Slow decomposition of coarse woody debris and a low rate of height growth of young spruces result in the long temporal gap of about 50 years between disturbance and establishment of new generation of spruce saplings. Because of the temporal gap between the disturbance and the seedling recruitment and a slow growth rate of young spruces, a lag of about 100 years between disturbance and establishment of 5-m tall young stand can be expected in the Carpathian subalpine forests.

Keywords: Carpathians Mountains, deadwood, field-layer vegetation, gap, mast seeding, Norway spruce, *Picea abies*, regeneration, subalpine forest.

INTRODUCTION

Tree regeneration depends on various relationships linking the establishment and development of young individuals with different element of the forest structure. The mortality pattern of canopy trees, i.e. the disturbance regime, is of particular importance for recruitment and growth of young individuals of tree species (Pickett and White 1985, Kuuluvainen 1994). Type, size, rate and severity of disturbances are decisive factors influencing conditions at the forest floor (Denslow 1987, Canham 1988a). In mixed forests, where many tree species represent various regeneration strategies any type of disturbance can provide suitable conditions for at least one species from the whole set. In contrast, in forests with a small number of tree species not all regeneration strategies are represented and the success of tree regeneration can be expected to depend on the specific conditions required by the dominant species, and not every disturbance fulfills such requirements. This simplification of species composition is especially pronounced in subalpine forests at high altitudes where only a limited number of tree species can exist in harsh climate, with low temperatures, strong winds and deep snow cover. The Carpathian subalpine forests are extremely poor in tree species as usually only two of them coexist at high elevations in these Central European mountains: *Picea abies* (L.) Karst., the only climax species, and *Sorbus aucuparia*, regarded as a pioneer. More tree species are found in spruce-dominated forests in the boreal zone, where three or four species coexist (Kuuluvainen 1994). Norway spruce is generally considered a shade-tolerant species (Ellenberg 1986), but in subalpine forests at high altitudes it becomes a light-demanding tree (Pisek and Winkler 1959). The relationship between spruce regeneration and gaps in subalpine forests can be expected to differ from that in the boreal forests of Scandinavia and Russia. This difference in the light requirements of young spruces between the high mountains of Central Europe and the boreal zone of Northern Europe should make the dynamics of the spruce forests in these two regions different as well.

Disturbances in tree stand contribute to the high spatial variability of understory vegetation, which is usually more exuberant in and near gaps than under closed tree canopy (Boone et al. 1988, Peterson and Pickett 1995, 2000). However, species composition does not necessarily change after gap formation and so called "gap specialists" are plants not restricted to gaps, but more frequent in gaps than under closed canopy (Goldblum 1997). Vegetation within gaps is not homogenous; there are differences between the central and peripheral parts of gaps, as well as variability caused by dead standing or knocked-down gap-makers (Wayne and Bazzaz 1993). It is well known that the dominance of different plant species in the field layer creates variable conditions for seed germination and development of young spruces (Reif and Przybilla 1995). The negative effect of bilberry *Vaccinium myrtillus* on spruce regeneration has been emphasized (Bernier and Ponge 1994, Jaederlund et al.

1996). For better understanding of the regeneration behavior of trees, the response of understory vegetation to gap disturbance should be considered also, so that the joint effect of both factors can be determined.

The establishment of specific microsites during disturbances and their later maintenance on the forest floor is also important for tree regeneration. A strong connection between young trees and the microhabitats created by the death of trees has been reported many times. Pioneers often settle on exposed mineral material of windthrow mounds (Putz 1983, Nakashizuka 1989, Yamamoto 1995). On the other hand, coarse woody debris is usually considered a suitable substrate for regeneration of conifers (Franklin and Hemstrom 1981, Lertzman 1992, Little et al. 1994). It is well known that coarse woody debris and windthrow mounds provide suitable conditions for regeneration of Norway spruce (Mayer et al. 1972, Jonsson 1990, Hörnberg et al. 1995, Reif and Przybilla 1995, Hofgaard 1993a, Zielonka and Niklasson 2001, Zielonka 2006a). Unfortunately, the data on tree regeneration are usually limited to the relative abundance of seedlings and saplings on different substrates. Little attention has been paid to what makes these substrates favorable for seed germination and development of young trees. In this respect, of special interest is the size and age of the logs and mounds invaded by young trees.

We report different aspects of spruce regeneration in Carpathian subalpine spruce forest. In particular we intend to present mutual relationship among field-layer heterogeneity, gap size, and variability of microsites created by death of trees and to reveal their joint effect on the establishment and development of a new generation of spruce. We intend to demonstrate whether and how the different field-layer species affect spruce regeneration; to find the relationship between spruce regeneration and the size and age of logs and windthrow mounds, structures created by the death of trees; to examine how the regeneration behavior of spruce is related to gap size; and to evaluate the role of different microhabitats in spruce regeneration.

STUDY AREA

The investigation was carried out in the massif of Babia Góra (1725 m), the highest area in the West Beskids (Polish Carpathians). Subalpine spruce forest occurs in this massif at 1150-1400 m a.s.l. The climate is cool at this elevation, with mean annual temperature of 2-4°C, mean annual rainfall of 1470 mm, snow depth of 1-2 m, and a snow-free period of 7 months

(Obrębska-Starklowa 1963, Holeksa and Parusel 1989). The soils are mainly humus-iron podzols, iron podzols and podzolized rankers that developed from magurian sandstone with thin mudstone interbreedings (Adamczyk 1989). The investigated spruce forest has been under protection since 1930. In 1955 it was incorporated into the strict reserve of the Babia Góra National Park and no management has been applied since that time.

All investigations were made in a 14.4 ha (340x424 m) tract located between 1190 and 1300 m on the northern slope of the massif of Babia Góra. The area includes a wide plateau of several hectares, an extensive area of gentle slopes 5-15°, and small fragments of steep slopes inclined up to 40°. The slopes are exposed mostly to the north and northeast. Over the whole area a 40x40 m grid consisting of 104 points was established with a theodolite and permanently marked in the field.

Spruce was the dominant species in the investigated forest, with 258 trees of dbh >10 cm per 1 ha. The basal area was 36 m^2/ha and the volume was 407 m^3/ha. The average diameter of these trees was 40 cm, and their average height 23 m. A sparse admixture of Sorbus aucuparia was present in the forest (Holeksa 2001).

There were 38 canopy gaps per 1 ha of the forest, constituting 33.8% of its area. Gap size ranged from 4 to 1473 m^2 (avg. 92 m^2). Most gaps were small (<50 m^2), and only a few were >500 m^2 (Holeksa and Cybulski 2001).

There were 112 logs of diameter >10 cm and length >3 m per 1 ha. Their volume was 72.6 m^3/ha. The area of downed trunks was 300 m^2/ha. Most logs represented classes I and IV of decomposition (Table 1), and those of diameter 21-30 cm covered the largest area. The area covered by logs was twice as large in gaps (451 m^2/ha) than outside gaps (226 m^2/ha). Logs within gaps were of greater diameter than those under the canopy (Holeksa 2001).

Nineteen types of patches in the field-layer vegetation were distinguished in the investigated forest (Holeksa 2003). They differed mainly in species abundance, with minor differences in species composition. Patches with high abundance of *Athyrium distentifolium* and *Vaccinium myrtillus* covered nearly 2/3 of the area. *A. distentifolium* patches were nearly twice as frequent in gaps as outside them. Patches with *Calamagrostis villosa* and *Rubus idaeus* also occurred more often within gaps than under the canopy. On the other hand, *Dryopteris dilatata, Polytrichum formosum* and *V. myrtillus* were significantly more abundant under tree crowns than in gaps (Holeksa 2003).

Table 1. Characteristics of logs in different classes of decomposition

Class of decomposition	Surface	Shape	Depth of penetration by sharp object	Location	Branches	Bark
I	smooth	round	wood is solid	supported by branches and highly elevated above ground	branches present	most often covers the whole log (partial lack of bark caused by bark beetles and woodpeckers)
II	smooth	round	surface only bends under the pressure of a sharp object	supported by thick boughs and elevated above ground	only thick boughs and their stubs present	tears off from bottom surface; lacks on snags
III	crevices several mm deep are present	round	to 1 cm	supported by bough stubs and slightly elevated above ground	only stubs of at least 2 cm thick present	occasionally present on upper surface
IV	crevices about 0,5 cm deep are present	round	to 3 cm	partly in contact with ground	only stubs at least 4-5 cm thick present	usually lack of any remnants
V	thick (several cm) pieces of wood tear off from the bottom surface; sides are cracked with crevices 1 cm deep	round	to 5 cm	in contact with ground along the whole length; elevated only over small depressions	lack of any remnants	lack of any remnants
VI	thick (several cm) pieces of wood tear off from sides	round	solid pieces of wood only in the central part of log	entirely adheres to the ground	lack of any remnants	lack of any remnants
VII	the whole log is covered with crevices several cm	distinctly flattened	through	entirely adheres to ground	lack of any remnants	lack of any remnants

Class of decomposition	Surface	Shape	Depth of penetration by sharp object	Location	Branches	Bark
	deep					
VIII	most often totally covered with mosses and vascular plants	creates long structure elevated above the ground	through	joined with ground	lack of any remnants	lack of any remnants

Observations of Spruce Regeneration

Three groups of young spruces were distinguished. The first was a cohort established in 1993 (cohort '93). This cohort resulted from mast seeding of spruce in 1992. The second consisted of seedlings at least two years old and shorter than 30 cm in 1993. The third, saplings, included spruces at least 30 cm tall and of dbh <10 cm.

Three main microhabitats were distinguished: (1) undisturbed soil covered with field-layer vegetation, (2) windthrow mounds, and (3) dead wood (logs and stumps). Windthrow pits were not considered, as they usually are very shallow in the subalpine spruce forest and are rather quickly covered by mosses and vascular plants, making it difficult to distinguish them from undisturbed soil.

Cohort '93 and seedlings. — Cohort '93 and seedlings were observed on five 40x40 m squares chosen randomly. Observations were made annually from 1993 to 1997 in the second half of September.

Individuals of cohort '93 growing on the soil were counted annually in 147 plots 1 m^2. In each of the five 40x40 m squares, between 22 and 30 such plots were in different types of field-layer patches (131 plots total) and 2-5 plots were on windthrow mounds (16 plots total). In patch types covering large areas the plots were placed at random, but in types covering small areas they were placed in every patch large enough to contain a plot. All individuals growing on logs situated within the five 40x40 m squares were counted every year.

Seedlings were counted annually in 243 plots of 1 m^2 in patches of field-layer vegetation and in 16 plots on windthrow mounds. The sample size for seedlings in the field-layer vegetation was larger than for cohort '93 because of the small number of seedlings (only 48 individuals grew in 243 plots). In 1993 all seedlings on logs (278 individuals), stumps (30) and windthrow mounds (53), and 180 seedlings growing within field-layer vegetation (48 from 243 plotsof 1 m^2 + 132 found additionally) were tagged with small plastic sticks. Their age was determined in 1993 through inspection of the morphology of their aboveground shoots. Survival of tagged seedlings was observed annually till 1997. Two additional measures were recorded for seedlings growing on logs: (1) diameter of the log at the rooting site and (2) log decomposition class.

The size of cohort '93 and the size of seedling bank for each year from 1993 to 1997 was calculated based on their density on different microsites in the five 40x40 m squares and the fraction of the forest area covered with the microsites. Areas of the different types of field-layer vegetation and the different classes of log decomposition over the whole investigated forest were known from previous measurements (Holeksa 2001, 2003). The area of windthrow mounds over the 14.4 ha forest track was determined based on their frequency in a grid of 1392 points spaced 10x10 m.

Saplings. — All individual saplings were mapped in the whole 14.4 ha tract. The 40x40 grid, logs, snags and gaps were used as points of reference in mapping the saplings. All the mentioned structures had been mapped earlier to 1:500 scale (Holeksa 2001, Holeksa and Cybulski 2001). The following data were recorded for saplings: (1) height of all individuals; (2) length of the last five internodes of individuals lower than 3 m; (3) microsite (soil, windthrow mound, log, stump); (4) type of field-layer vegetation (for saplings on soil); (5) prevailing class of log decomposition (for saplings on logs); (6) diameter of log at rooting site (for saplings on logs); (7) class of decomposition and dbh of windthrow trees (for saplings on mounds); (8) size of gap (for saplings in gaps); and (9) distance to the nearest gap and its area (for saplings outside gaps). Data on the spatial relationships between saplings, young trees and gaps were read from the 1:500 map of their positions.

Statistical Analyses

Cohort '93 and seedlings. — The significance of differences in the density of cohort '93 and seedlings between field-layer patches was analyzed with ANOVA or the nonparametric Kruskal-Wallis test, depending on the distribution of data.

The distribution of cohort '93 and seedlings on logs was analyzed with the χ^2 test. It was assumed that the distribution of individuals was independent and that under random conditions the number of individuals on logs from different decomposition classes and 10-cm diameter classes were proportional to their area. The area of log sections falling into 10-cm diameter classes (10-20 cm, 20-30 cm…) was known from previous investigations (Holeksa 2001).

To determine the survivorship of cohort '93 in different field-layer types and on windthrow mounds, the percentage of individuals surviving from 1993 was calculated for every 1 m^2 plot. The differences between field-layer types were then examined with the Kruskal-Wallis test.

The significance of differences in survivorship of seedlings growing in different microhabitats was checked with the χ^2 test. The distributions of individuals in 1993 were compared with the distributions of individuals alive in subsequent years.

Saplings. — The χ^2 test was used to analyze the spatial distribution of saplings against the distribution of microhabitats (field-layer types, gaps of different sizes, windthrow mounds of different ages and sizes, logs of different diameters and ages). It was assumed that under random conditions the number of saplings growing in different microhabitats is proportional to the area covered by these microhabitats. Only in the case of mounds was their number instead of the area considered. Differences in sapling height between microhabitats were analyzed with ANOVA after logarithmic transformation of the height values.

The spatial pattern of saplings was examined with Ripley's K function (Diggle 1983). Sapling-to-sapling distances were used and the analysis was performed with the software SPPA (Spatial Point Pattern Analysis) version 2.0.3 created by Peter Haase (Haase 1995).

RESULTS

Cohort '93

Cohort '93 on soil and on windthrow mounds. — The number of germinants on soil in 1993 varied from 0 to 165 in the 1 m^2 plots. The highest density (nearly 80 individuals/m^2) was noted in patches dominated by *Dryopteris dilatata* and with two dominants *Vaccinium myrtillus* + *D. dilatata* (Figure 1). Density was also high, from 40 to 60 individuals per 1 m^2, in three field-layer patch types, dominated by *V. myrtillus*, *Calamagrostis villosa* or

Luzula sylvatica. Fewer than 20 germinants/m^2 were recruited within patches of *Deschampsia flexuosa* and *V. myrtillus* + *Polytrichum formosum*, as well as on windthrow mounds.

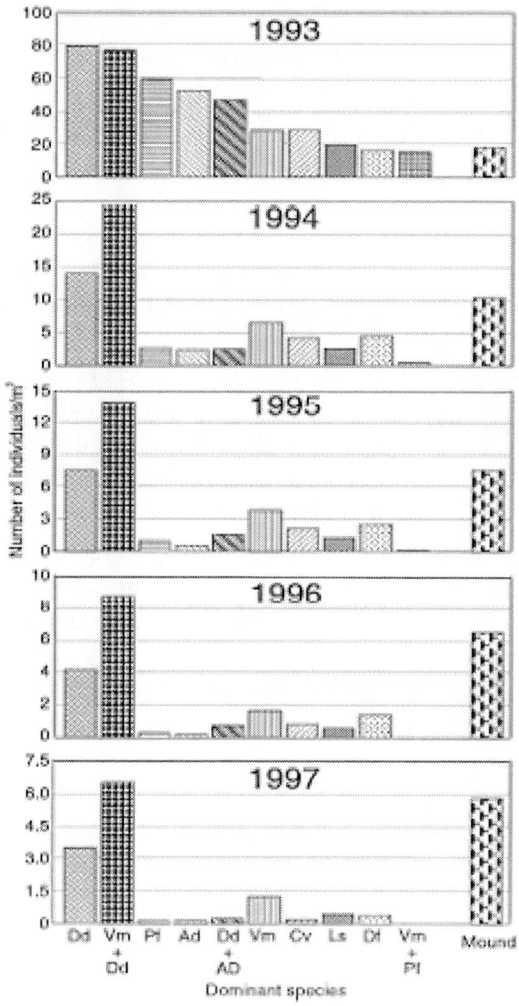

Figure 1. Changes in the density of spruce cohort '93 in field-layer patches and on windthrow mounds in 1993-1997. Density is expressed as the number of individuals per 1 m^2 of patches and per 1 m^2 of mound surface. Note that different density scales are used for consecutive years. Abbreviations of dominants: Ad – *Athyrium distentifolium*; Cv – *Calamagrostis villosa*; Dd – *Dryopteris dilatata*; Df – *Deschampsia flexuosa*; Ls – *Luzula sylvatica*; Pf – *Polytrichum formosum*; Vm – *Vaccinium myrtillus*.

The differences in the density of germinants in 1993 among the field-layer patch types and windthrow mounds were significant (ANOVA: $F_{10,\ 123} = 6.74$, $p < 0.001$). In 1994, the density of two-year old individuals was again highest in patches dominated by *D. dilatata* and *V. myrtillus* + *D. dilatata*, but was about a fourth of that in the previous year. The lowest density, below 1 individual/m^2, was noted in patches with *V. myrtillus* + *P. formosum* as dominants In 1997 the cohort was five years old and the highest density was on windthrow mounds and in patches of *D. dilatata* and *V. myrtillus* + *D. dilatata* (Figure 1). The lowest density was under *Athyrium distentifolium* and in patches with high cover of *P. formosum*. Differences in the density of individuals from cohort '93 among the field-layer patch types and mounds were significant for every year from 1994 to 1997 (Kruskal-Wallis test: $p < 0.001$ for all years).

The distribution patterns of cohort '93 differed between 1993 and 1994. This is emphasized by the low similarity in the ranking of field-layer patch types by cohort density (Spearman rank correlation: $r = 0.24$, $p = 0.054$). In subsequent years the pattern changed only slightly and the similarities between consecutive years in the ranking of field-layer patch types were high (r was from 0.88 to 1.0, and $p < 0.001$ for all comparisons).

The survivorship of cohort '93 was highest on windthrow mounds: more than 30% of the germinants from 1993 reached age five years (Figure 2). From 4% to 8% survived five years in patches dominated by *V. myrtillus* and/or *D. dilatata*. Still lower survivorship was noted in patches with *L. sylvatica*, *D. flexuosa*, *P. formosum*, *A. distentifolium* and *C. villosa*: from below 1% to 2%. Differences in four-year survivorship between field-layer patch types were significant (Kruskal-Wallis test: $p < 0.001$).

Cohort '93 on logs. — In 1993, germinants were on logs representing all classes of decomposition. Their density increased with the progress of log decomposition (Figure 3). On dead wood in class I the density was below 5 individuals/m^2. It was four times higher on logs in decomposition class II. On logs in classes III–VII the densities of germinants were similar, between 20 and 40 individuals/m^2. Still higher density, 80 germinants/m^2, was noted on logs in class VIII. The density of cohort '93 decreased gradually in 1994-1997, but the ratios between densities on logs of different classes of decomposition remained unchanged till 1997. The distribution of cohort '93 was highly correlated with the log decomposition class in the whole four-year period of observations ($\chi^2 = 170 \div 1424$, $df = 7$, $p < 0.001$ for all years).

In 1993 the average density of cohort '93 on soil was nearly 50 individuals/m^2, 2.5 times higher than the average density on logs (Figure 4).

Only on the most decomposed logs in class VIII, did the density exceed 40 individuals/m^2, comparable with the values noted on soil (comp. Figs 1 and 3). In subsequent years the proportions became reversed, and in 1997, when cohort '93 was 5 years old, in only two types of field-layer vegetation did the densities match those noted on logs. Even on the least decomposed wood, representing class I, the number of five-year-old spruces was higher than in most field-layer patch types.

The four-year survivorship of cohort '93 was not correlated with log decomposition ($\chi^2 = 9.4$, df = 7, p = 0.23). It varied from 12% on logs of class I to 21% on logs of class VII (avg. 18%). These figures were lower than on windthrow mounds, but much higher than on soil covered with different vegetation types (Figure 2).

Size of cohort '93. — The number of germinants in 1993 was nearly half a million per 1 ha. Of this huge number, only 7,200 grew on logs, stumps and windthrow mounds (Figure 5). These microsites occupied 3.9% of the total forest area. At the end of the second year of life, the size of cohort '93 diminished by a factor of seven, and 5.1% of the individuals grew on microsites created by the death of trees. By 1997, cohort '93 was reduced to 13,100 individuals per 1 ha and 11.8% of them grew on dead wood and mounds.

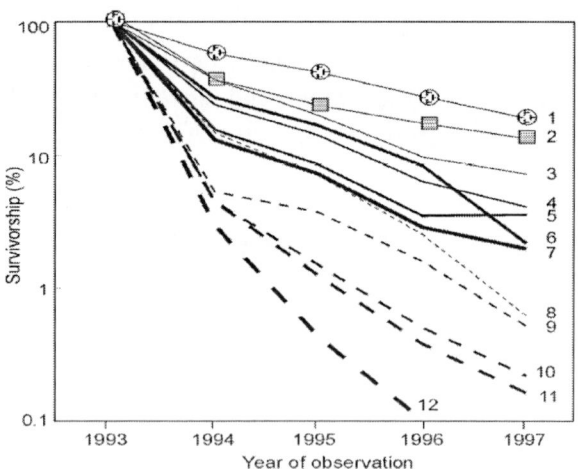

Figure 2. Survivorship of spruce cohort '93 in 1993-1997: (1) on windthrow mounds, (2) on logs of all decomposition classes combined, and in field-layer patches dominated by: (3) *Vaccinium myrtillus* + *Dryopteris dilatata*, (4) *Vaccinium myrtillus*, (5) *Dryopteris dilatata*, (6) *Deschampsia flexuosa*, (7) *Luzula sylvatica*, (8) *Calamagrostis villosa*, (9) *Dryopteris dilatata* + *Athyrium distentifolium*, (10)

Polytrichum formosum, (11) *Athyrium distentifolium*, (12) *Vaccinium myrtillus* + *Polytrichum formosum*.

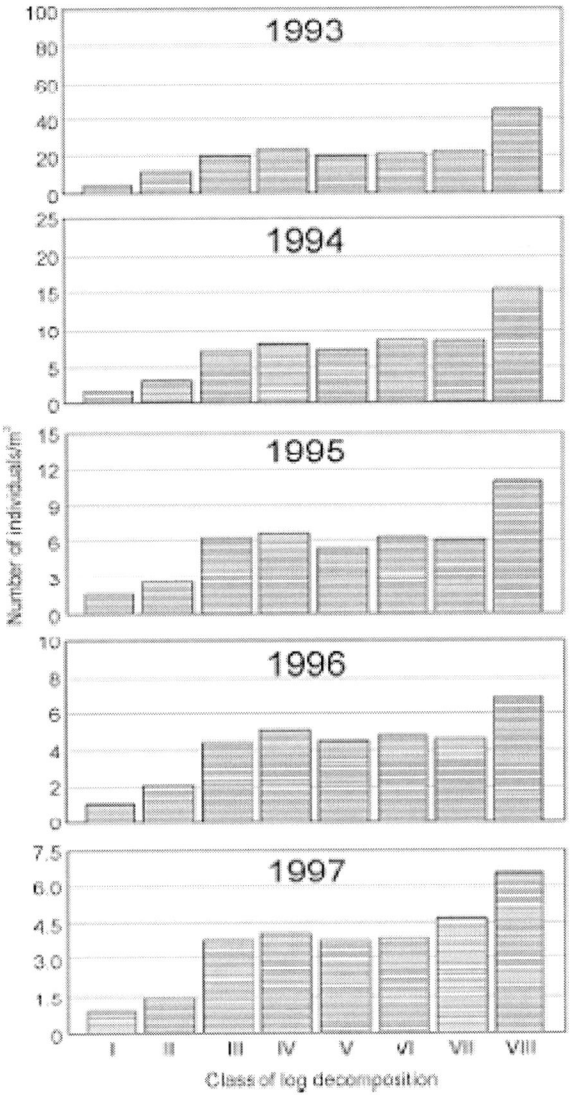

Figure 3. Changes in the density of spruce cohort '93 on logs in 1993-1997, by decomposition class. Density is expressed as the number of individuals per 1 m^2 of log surface. Note that different density scales are used for consecutive years.

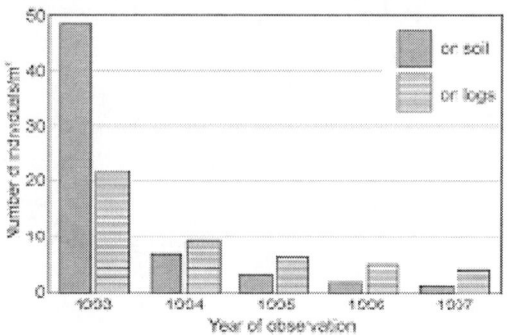

Figure 4. Changes in the density of spruce cohort '93 on soil and on logs. All types of field-layer vegetation as well as all classes of log decomposition are summed. Density is expressed as the number of individuals per 1 m² of substrate.

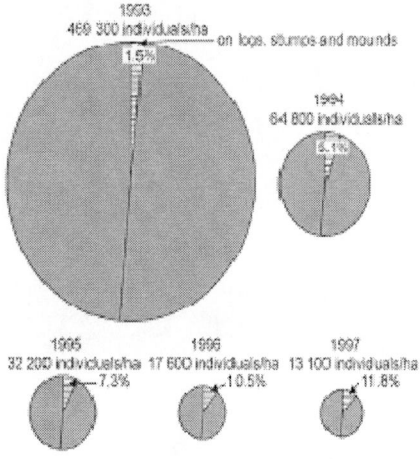

Figure 5. Changes in the size of spruce cohort '93 and in the percentage of spruces growing on logs, stumps and windthrow mounds in 1993-1997. Cohort size is expressed as the number of individuals per 1 ha of forest.

Seedlings

Seedlings on soil and windthrow mounds. — The density of seedlings ranged from 0 to 0.7 individuals/m² in different field-layer patch types (Figure 6), and was 4.5 individuals/m² on windthrow mounds. There were no seedlings within patches dominated by *A. distentifolium*, *L. sylvatica* and *P. formosum*.

Their density was highest in patches with high cover of *V. myrtillus*. The differences in seedling density in 1993 among field-layer patch types and windthrow mounds were significant (Kruskal-Wallis test: p <0.001).

Of 180 seedlings growing on soil tagged in 1993, 113 grew within concentrations of *V. myrtillus* and from 1 to 15 in the remaining patches. Also tagged were 53 seedlings on windthrow mounds. The tagged individuals were observed till 1997. Four-year survivorship was 63% in patches of *V. myrtillus*, 73% in other field-layer types altogether, and 75% on mounds; the differences were not significant ($\chi^2 = 1.23$, df = 2, p = 0.54).

Seedlings on dead wood. — The density of several-year-old spruces correlated with the log decomposition class ($\chi^2 = 70.9$, df = 6, p <0.001). Seedling density in 1993 varied from 1 individual/20 m^2 on class I logs to 2 individuals/m^2 on class VII logs (Figure 7A).

The occurrence of spruce seedlings also correlated with log diameter ($\chi^2 = 77.0$, df = 3, p <0.001; logs in class I were not considered in this analysis as only 2 seedlings were found on them). The thicker the logs, the higher the density of the seedlings on them, increasing from 0.5 individuals/m^2 on logs thinner than 20 cm to 2.2 individuals/m^2 on logs thicker than 50 cm (Figure 7B).

About 50% of the seedlings survived from 1993 to 1997 on logs in classes II, III and VIII of decomposition (Figure 8A). On logs in classes IV to VII the survivorship was from 70% to 90%. These differences were nearly significant ($\chi^2 = 12.3$, df = 6, p = 0.06). Seedling survivorship also correlated with log diameter (Figure 8B). It was significantly lower on logs thinner than 20 cm ($\chi^2 = 3.9$, df = 1, p = 0.05).

A number of seedlings grew on stumps up to 1 m high: 0.7 individuals per 1 m^2 of stumps in 1993, decreasing to 0.5 individuals/m^2 in 1997.

The survivorship of seedlings growing on dead wood was 71% and did not differ significantly from the survivorship of seedlings growing on soil ($\chi^2 = 0.19$, df = 1, p = 0.66).

Size of seedling bank. — The size of the seedling bank was 2040 individuals/ha in 1993. It diminished to 1370 individuals/ha in 1997. The percentage of seedlings growing on logs, stumps and windthrow mounds was 15.2% in 1993 and 16.1% in 1997.

Age structure of seedlings. — In 1993, 541 seedlings were tagged. The age of seedlings growing on various microsites did not differ significantly (Kolmogorov-Smirnov test, p >0.05 in all comparisons). The most numerous were five-year-old seedlings established in 1989 (Figure 9). Most individuals determined as three and four years old probably had in fact reached the age of five years. Their ages had been determined incorrectly because the annual leader increments were poorly distinguished during field observations, when they had to be handled carefully to avoid injuries that could lower seedling survival.

Survivorship of 3-5-year-old seedlings and older ones did not differ significantly. For both groups it was about 70% ($\chi^2 = 2.69$, df = 1, p = 0.1). As a result, the age structure did not change during four years, and in 1997 the seedlings established in 1989 still markedly outnumbered other age classes.

Saplings

General characteristics of sapling population. — There were 1281 spruces >30 cm high and with dbh <10 cm. Their density was 92 individuals/ha. Nearly 50% of them were shorter than 90 cm. Saplings a few meters high were rare; only 9% of them exceeded 3 m (Figure 10). Saplings grew in clusters in the whole analyzed range of spatial scales (Figure 11A).

Saplings on soil. — There were 481 (38%) saplings growing on soil. Their density was 35 individuals/ha. They were distributed in clusters for all distances from 2 to 50 m (Figure 11B).

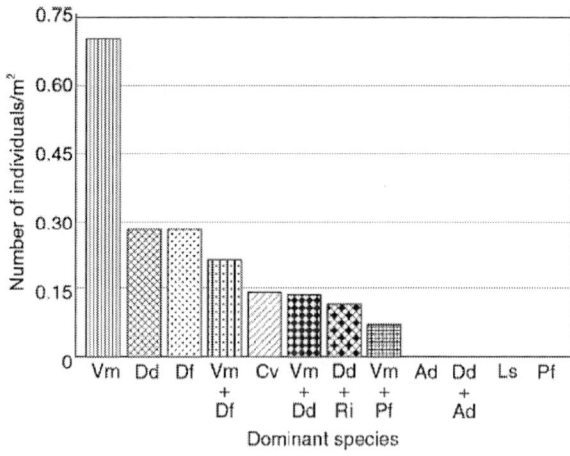

Figure 6. Density of seedlings in 1993 in field-layer patches. Density is expressed as the number of individuals per 1 m² of patch. Abbreviations of dominants: Ad – *Athyrium distentifolium*; Cv – *Calamagrostis villosa*; Dd – *Dryopteris dilatata*; Df – *Deschampsia flexuosa*; Ls – *Luzula sylvatica*; Pf – *Polytrichum formosum*; Ri – *Rubus idaeus*; Vm – *Vaccinium myrtillus*.

Norway Spruce *Picea Abies* Regeneration and Canopy Disturbance ... 17

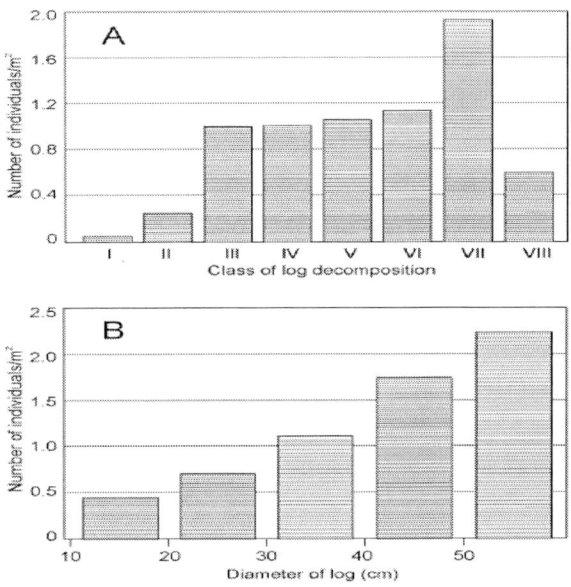

Figure 7. Density of seedlings in 1993, by (A) log decomposition class and (B) log diameter class. Density is expressed as the number of individuals per 1 m^2 of log surface.

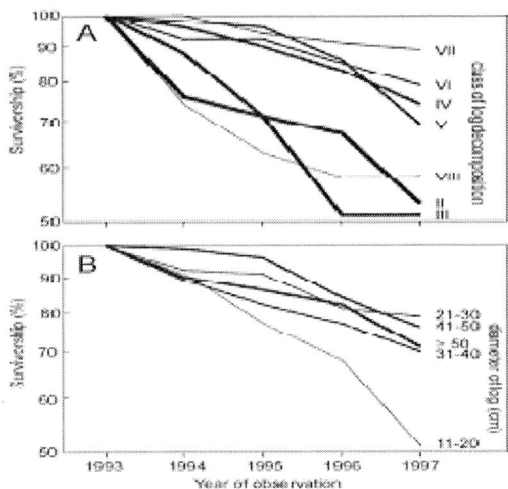

Figure 8. Survivorship of seedlings growing on logs in 1993-1997 versus (A) log decomposition class and (B) log diameter.

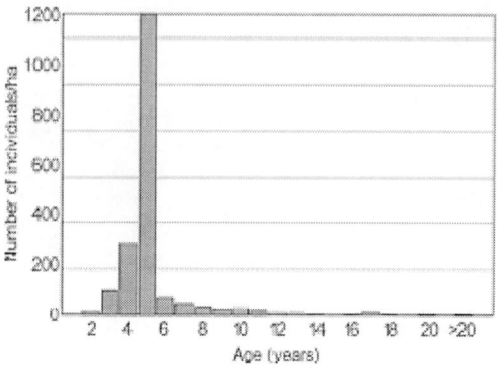

Figure 9. Age structure of seedlings in 1993. Only individuals at least 2 years old in 1993 are considered. Number of individuals in each age class is expressed per 1 ha of forest.

The distribution of saplings correlated with the type of field-layer vegetation ($\chi^2 = 480$, df = 10, p <0.001). Their highest density was in patches dominated by *D. dilatata* + *R. idaeus* (Figure 12A) occupying only 0.8% of the forest, and their contribution to the whole sapling population was small (Figure 12B and C). In three other types dominated by *V. myrtillus* and/or *D. dilatata* the density was from 0.3 to 1.1 individuals per 100 m^2. These patches covered 34% of the forest area and their contribution to spruce regeneration was relatively high: 75% of the saplings on soil grew in patches of these field-layer types. The lowest density of saplings was in patches dominated by *A. distentifolium*, and in spite of the large area covered with the fern (42% of the investigated area) only a small share (3.5%) of the saplings growing on soil were within its dense conglomerations.

Height structure was analyzed for only six field layer patch types in which more than 10 saplings were found. Saplings in patches dominated by *V. myrtillus* were lower than those in the others (Figure 13). On average they were 30 cm lower than saplings in patches with two dominants: *V. myrtillus* + *P. formosum* and *D. dilatata* + *R. idaeus*. The tallest saplings were accompanied by *D. dilatata* alone and *D. dilatata* + *V. myrtillus*. Sapling height differed significantly between these six types (ANOVA: $F_{5, 414} = 10.6$, p <0.001; values were log-transformed).

Saplings on windthrow mounds. — On mounds, 165 saplings (13%) were found. Their density was 18 individuals per 100 m^2 of this microsite and 12 individuals per one hectare of forest. The saplings on mounds were distributed in clusters over the whole analyzed range of spatial scales (Figure 11C).

Clumping was higher at two distances, the first at a few meters and the second at 10-15 m between saplings. These scales correspond with clusters on individual mounds and on small groups of mounds.

The spatial distribution of saplings correlated with mound age defined by the decomposition class of the accompanying logs ($\chi^2 = 37.8$, df = 7, p <0.001). There were only a few individuals on the youngest mounds with logs in classes I and II of decomposition, one sapling for every 30 mounds. The density of saplings on older mounds was much higher, about one individual per 2 mounds (Figure 14A). A number of saplings grew on mounds that were not accompanied by logs. These mounds were ranked as class >VIII of decomposition.

The distribution of saplings was also correlated with mound size defined by the diameter of accompanying logs ($\chi^2 = 109$, df = 4, p <0.001). This analysis was limited to decomposition classes III-VII because of the scarcity of saplings on the youngest and oldest mounds. The density of saplings, measured as the number of individuals per mound, increased with mound size. There was one sapling for every 26 mounds accompanied by logs of diameter <20 cm, and nearly 2 saplings for every mound created by the fall of trees >50cm (Figure 14B).

Mounds linked to logs of dbh <20 cm were of little importance for spruce regeneration, as only 3% of the saplings grew on them in spite of their large number (Figure 14C). Windthrows of trees 30-40 cm, which were as numerous as the smallest, were the most important for spruce regeneration. They supported nearly 40% of all saplings growing on mounds. Every sixth spruce grew on mounds created by the fall of trees thicker than 50 cm, and such trees comprised only 5% of all uprooted trees.

The mean height of saplings on mounds was 101 cm. Most of them were shorter than 70 cm. Sapling height did not vary significantly between mounds of different ages (ANOVA: $F_{4, 137} = 0.52$, $p = 0.72$; values were log-transformed; only mounds in classes III-VII were considered).

Saplings on dead wood. — On the whole investigated area, 633 saplings (49%) were found on logs and stumps. Their density was 12.5 individuals per 100 m^2 of dead wood and 45 individuals per hectare of forest. Saplings on logs and stumps grew in clusters over the whole range of analyzed distances (Figure 11D). Clumping was higher at distances of a few meters, as indicated by the inflection near the beginning of the graph of function $L(t)$. This probably corresponds to small clumps on individual logs.

The distribution of spruce saplings was related to log age ($\chi^2 = 566$, df = 7, p <0.001). No saplings were found on logs in decomposition classes I and II

(Figure 14D). On logs representing classes III-VII the density of saplings increased with the degree of decay from 5 to 31 individuals per 100 m^2. The differences were especially pronounced between classes IV and V. Nearly 20% of the saplings growing on deadwood were rooted on wood remnants not visible on the soil surface. This very decomposed woody debris was ranked as class >VIII of decomposition.

The distribution of saplings also correlated with log thickness ($\chi^2 = 271$, df = 5, p <0.001; only logs in classes III–VIII of decomposition were considered). There were 3.2 saplings per 100 m^2 of logs thinner than 20 cm (Figure 14E). Density increased with log diameter, reaching 47 individuals/100 m^2 on those thicker than 60 cm.

In the next analysis, regeneration was considered versus dbh of trees that became logs. Only logs in classes III–VI of decomposition were included, because logs in classes VII–VIII were fragmented and the breast height position could not be determined exactly. Saplings were very rare on logs of dbh <30 cm (Figure 14F). One spruce grew per 21 logs of dbh <20 cm, and on average almost 2 saplings per log of dbh >60 cm.

The role of lying trunks of different dbh in spruce regeneration depends on their suitability and abundance. Only 4% of the saplings grew on downed trunks with dbh <20 cm (Figure 14G), even though these small logs constituted as much as 29% of the number and 24% of the area of all logs. Trunks of dbh >60 cm constituted only 4% of both the number and area of all logs, but supported 18% of the saplings. Downed trees with dbh between 30 and 50 cm were the most important for regeneration (Figure 14G).

Sapling height differed between consecutive classes of log decomposition (ANOVA: $F_{6, 540} = 29.3$, p <0.001; values were log-transformed). The average height ranged from 56 cm on class III logs to 172 cm on class VIII logs, and 179 cm on still more decomposed wood (class >VIII) (Figure 15). The height of spruces growing on logs in classes III-V differed slightly. Greater differences were noted between spruces on logs of the next classes. Saplings lower than 70 cm prevailed on logs up to class VII. Nevertheless, individuals higher than 2 m also formed quite a numerous group on logs in class VI and beyond (Figure 15).

Saplings in gaps and under spruce canopy. — Sapling distribution strongly correlated with gaps ($\chi^2 = 815$, df = 1, p <0.001). As much as 73% of the saplings (930 individuals out of 1281) grew in gaps, which occupied 34% of the forest area. Sapling density was 192 individuals/ha in gaps and 38 individuals/ha outside gaps.

The distribution of saplings correlated with gap size ($\chi^2 = 387$, df = 4, p <0.001). Their density increased from 0.7 individuals/100 m^2 in gaps smaller than 50 m^2 to 6 individuals/100 m^2 in gaps larger than 800 m^2 (Figure 16A). There were no saplings at all in most small gaps; they were present in only 10% of gaps <100 m^2.

Saplings outside gaps tended to grow close to them ($\chi^2 = 60.9$, df = 8, p <0.001). Their density was 0.6 individuals/100 m^2 near canopy gaps (up to 2 m from them), dropping below 0.1 individuals/100 m^2 farther than 6 m from gaps (Figure 16A). There were more saplings around large than around small gaps. Their density in a 4 m wide strip increased from 0.2 individuals/100 m^2 around gaps <50 m^2 to 2.4 individuals/100 m^2 in the vicinity of gaps >800 m^2 (Figure 16B).

The mean height of saplings in gaps and under spruce canopy was similar (126 and 129 cm, respectively) and did not differ significantly (ANOVA: $F_{1, 1276} = 0.06$, p = 0.81; values were log-transformed).

Saplings on different microsites – comparison. — The distribution of saplings was strongly correlated with microsites ($\chi^2 = 10012$, df = 3, p <0.001). More than 50% of the spruces grew on visible remnants of trees (logs and stumps) or on windthrow mounds, which occupied 3.3% and 0.6% of the forest area, respectively (Figure 17). Very decomposed pieces of wood were dug out from between the roots of 10% of the saplings. A third of the saplings grew on undisturbed soil covered with field-layer vegetation. One spruce sapling grew per 300 m^2 of soil, per 9 m^2 of visible woody debris, and per 4 m^2 of windthrow mounds.

Saplings on logs, stumps and mounds were more linked to gaps than were saplings growing on soil ($\chi^2 = 74.6$, df = 1, p <0.001): 92% and 78% of the saplings growing on mounds and decomposed wood, respectively, were situated in gaps, while only 58% those growing on soil were in gaps. As a result, in gaps the density of saplings growing on logs, stumps and mounds was more than twice the density of saplings growing on soil (1.3 vs. 0.6 individuals per 100 m^2 of gaps), while under spruce canopies the ratio was more or less even, with more saplings on soil (0.17 vs. 0.22 individuals per 100 m^2 of stand understory).

The average height of saplings differed between the three microsites (ANOVA: $F_{2, 1276} = 4.6$, p = 0.01; values were log-transformed). The highest saplings grew on soil (avg. 137 cm). Saplings on logs and stumps were slightly shorter (121 cm), and those on windthrow mounds were the shortest (101 cm). The height of saplings growing on particular microsites did not differ significantly between inside-gap and outside-gap locations.

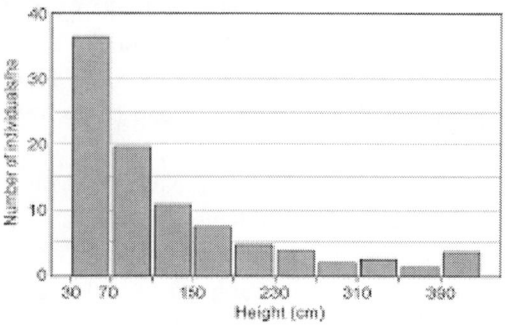

Figure 10. Height distribution of saplings. Number of individuals in each height class is expressed per 1 ha of forest.

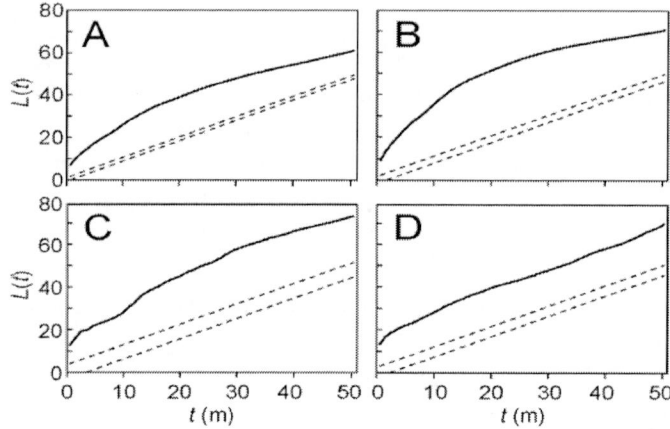

Figure 11. Spatial pattern of saplings. Results of analyses are shown for (A) all saplings, (B) saplings growing on soil, (C) saplings growing on logs and (D) saplings growing on windthrow mounds. Dashed lines delimit the area of random distribution; the area of clumped distribution is above them; solid line indicates function L(t).

Sapling height was not correlated with mound age defined as the decomposition class of accompanying logs, while the height of saplings on logs increased with increasing decomposition (Figure 15). As a result, for decomposition classes III and IV the saplings on mounds were higher than those on logs, while for classes V and VI the differences were small and not significant (ANOVA: class III – $F_{1, 36} = 11.0$, $p = 0.002$; class IV – $F_{1, 114} = 7.8$, $p = 0.006$; class V – $F_{1, 121} = 1.6$, $p = 0.21$; class V – $F_{1, 168} = 0.01$, $p = 0.92$; values were log-transformed).

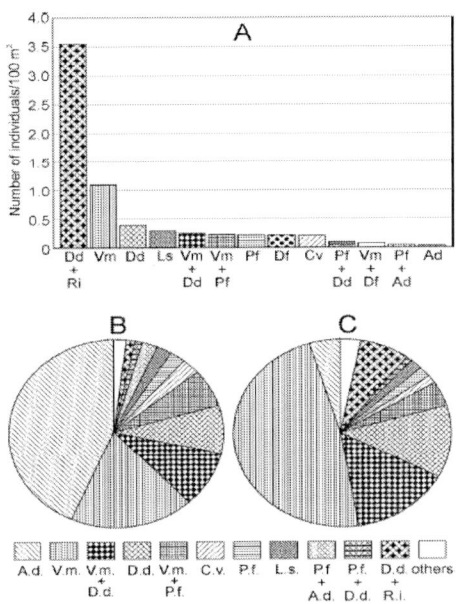

Figure 12. (A) Density of saplings, by field-layer patch type. (B) Percentage of forest area covered with different patch types. (C) Distribution of saplings, by patch type. Density in figure A is expressed as the number of individuals per 100 m² of patch. Abbreviations of dominants: Ad – *Athyrium distentifolium*; Cv – *Calamagrostis villosa*; Dd – *Dryopteris dilatata*; Df – *Deschampsia flexuosa*; Ls – *Luzula sylvatica*; Pf – *Polytrichum formosum*; Ri – *Rubus idaeus*; Vm – *Vaccinium myrtillus*.

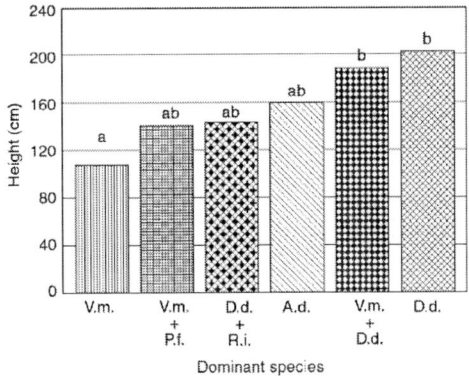

Figure 13. Average height of saplings in field-layer patches. Different letters above bars denote significant differences (Tukey test, $p < 0.05$). Abbreviations of dominants: Ad – *Athyrium distentifolium*; Dd – *Dryopteris dilatata*; Pf – *Polytrichum formosum*; Ri – *Rubus idaeus*; Vm – *Vaccinium myrtillus*.

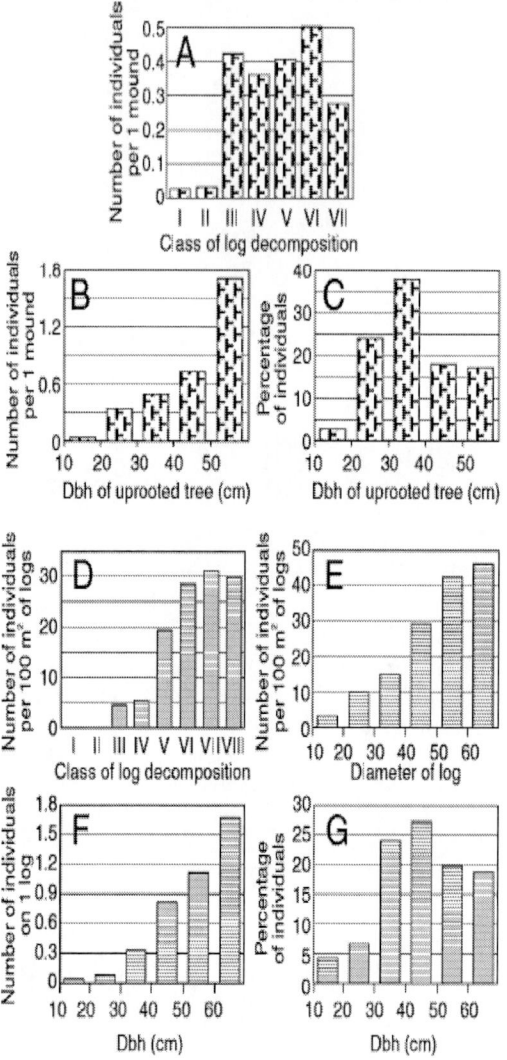

Figure 14. Density of saplings on windthrow mounds and logs. Density of saplings on mounds, by (A) decomposition class of accompanying log and (B) dbh of accompanying log. (C) Distribution of saplings on mounds, by dbh of accompanying log. Density of saplings, by (D) log decomposition class, (E) log diameter class, (F) log dbh. (G) Distribution of saplings, by log dbh. Density is expressed as the number of individuals per mound or per 100 m^2 of log surface.

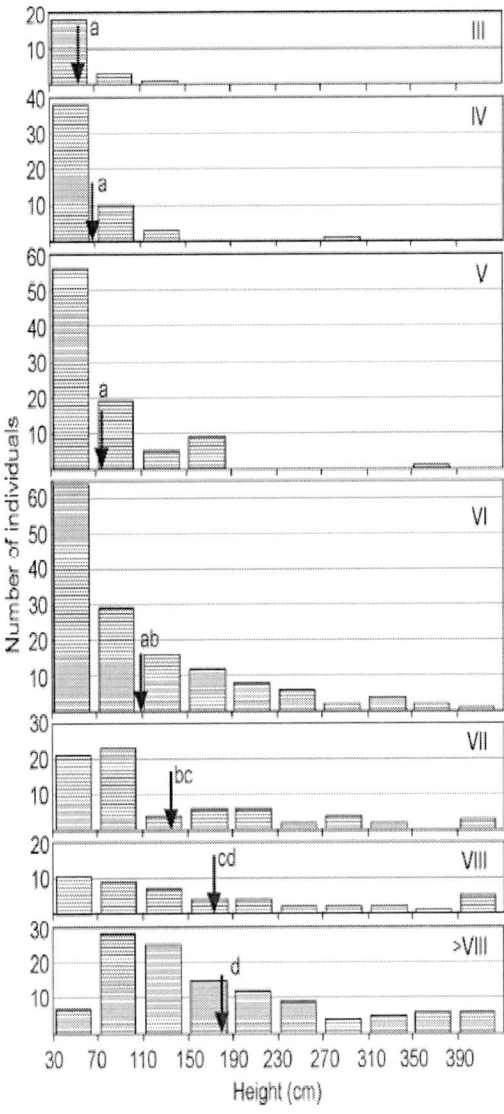

Figure 15. Height distribution of saplings growing on logs, by decomposition class. Arrows point to average heights of saplings. Different letters by arrows indicate significant differences (Tukey test, $p < 0.05$). Number of individuals refers to the area of 14.4 ha. Decomposition class >VIII refers to heavily decomposed wood visible only if the root system of the sapling was dug.

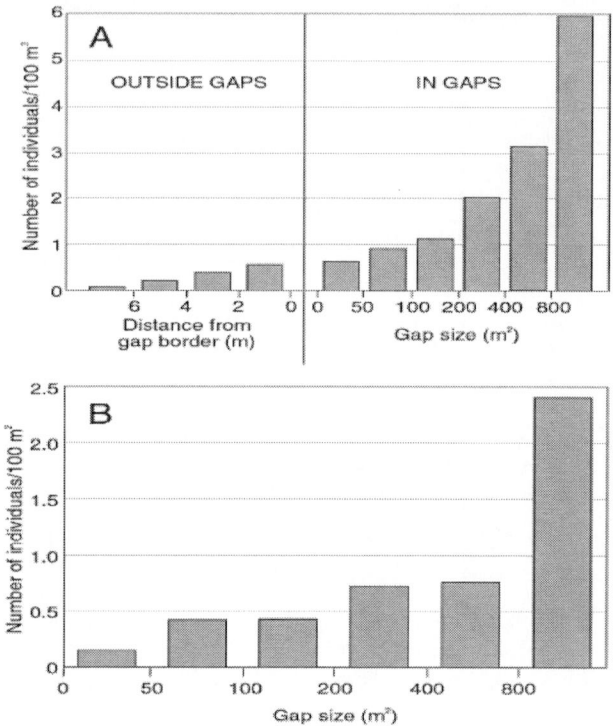

Figure 16. (A) Density of saplings at different distances from gaps and in gaps of different sizes. (B) Density of saplings in 4 m wide belts surrounding gaps of different sizes. Density is expressed as the number of individuals per 100 m^2 of gaps and spruce understory.

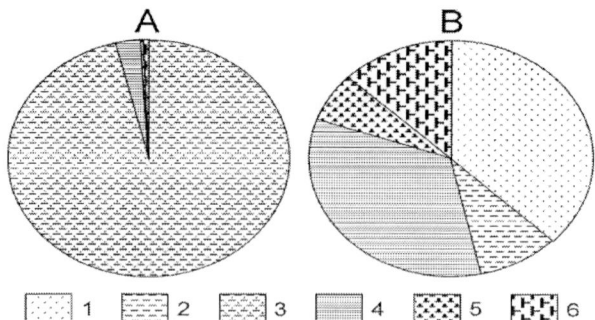

Figure 17. (A) Percentage of area covered with different substrates. (B) Distribution of saplings, by substrate. (1) soil, (2) highly decayed and invisible woody debris, (3) soil and highly decayed and invisible woody debris, (4) logs, (5) stumps, (6) windthrow mounds.

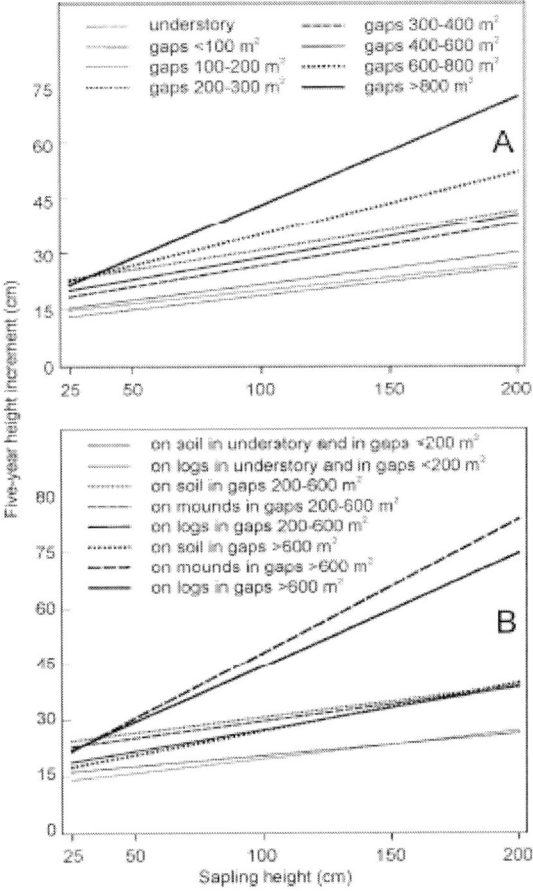

Figure 18. Relationship between sapling height in 1988 and average annual height increment during the next five years, given for (A) all saplings in the understory and in gaps of different sizes, (B) saplings on soil, mounds and logs situated in the understory and in gaps of different sizes.

The rates of growth in sapling height differed between those in gaps and under spruce canopy, as well as between microsite types (only saplings with undamaged tops that were 25-200 cm high were considered). In the understory and in small gaps <200 m^2 the growth rates were similar (Figure 18A). Linear regressions indicated that spruces 25 cm high grew 15 cm under stand canopy and 15.5 cm in gaps 100-200 m^2 in five years. The figures for 2-m saplings were 27.1 and 30.3 cm, respectively. The growth of saplings in large gaps >800 m^2 was much quicker. Small (25 cm) spruces grew 21.6 cm in five years, nearly doubling their height, and those 2 m high grew 71.7 cm.

Under spruce canopies and in gaps up to 600 m² the growth rates of saplings were similar between microsite types. The differences between the microsites were pronounced only in large gaps >600 m² (Figure 18B): 25 cm saplings there grew about 20 cm in five years irrespective of the microsite type. The differences were much greater for saplings 2 m high. On mounds and downed wood, individuals of that height grew about 80 cm in five years, and on soil only 40 cm. The growth rates of saplings on soil were similar in gaps 200-600 m² and >600 m², but saplings on mounds and deadwood grew much quicker in large gaps >600 m² than in smaller gaps.

DISCUSSION

The Role of Field-Layer Vegetation in Spruce Regeneration

The patch approach used here to analyze the interactions between field-layer vegetation and spruce regeneration is similar to that applied in other studies of tree regeneration (Burton and Bazzaz 1995). It is also consistent with the synusial approach of Barkman (1973). According to Barkman (1973), both patches and synusiae can be defined as elementary one-layered vegetation units that are floristically, physiognomically and ecologically homogenous, and associated with uniform environmental conditions.

The input of spruce seeds to the soil can be assumed to be independent of the field-layer vegetation. Thus the observed differences in germinants and seedling density were caused by an environmental sieve formed by the heterogeneity of dominant plants. Up to now, particular attention has been paid to the response of spruce regeneration to the negative influence of the thick organic layer found under *Vaccinium myrtillus* (Pellissier and Trosset 1989, Jaederlund et al. 1996, Mallik and Pellissier 2000). Poor spruce regeneration in patches dominated by *V. myrtillus* was also noted in the Bavarian Forest in Germany (Reif and Przybilla 1995), but its negative influence was weaker than that of other species. On the contrary, Rehak (1957) stated that spruce regeneration was good where the ground was covered with *V. myrtillus* or mosses. The results from the Babia Góra subalpine forest fully support those from the Bavarian Forest and Czech. Seedling density, sapling density and seedling survivorship were highest in patches dominated by *V. myrtillus*. The disparity between the results of experimental studies (Pellissier and Trosset 1989, Jaederlund et al. 1996, Mallik and Pellissier 2000) and field observations (Reif and Przybilla 1995, this study) is only apparent. It should be stressed that a density of one sapling per 100 m² of patches with *V. myrtillus* is far too low to establish a dense new stand. Moreover, bilberry cover is not continuous, and spruce seedlings can settle in the many small gaps found among thickets where litter deposition is less intensive. Such gaps do not exist within patches of another widespread dominant, *Athyrium distentifolium*. Probably a number of saplings growing within the remaining types were established within conglomerations of bilberry, and this species gave way to other field-layer plants only after gaps opened. Such a possibility can be inferred from the results of Parusel and Holeksa (1991) and Holeksa (2003).

In patches of *Athyrium distentifolium* and *Dryopteris dilatata* the density of germinants was higher than in all others except patches with *Polytrichum formosum*. Soil conditions, especially thin organic horizons due to quick decomposition of litter (Adamczyk 1989), are probably responsible for good seed germination under ferns. They are usually situated in canopy gaps (Parusel and Holeksa 1991, Holeksa 2003) where deposition of spruce litter is less than under tree crowns. Dense conglomerations of *A. distentifolium* occupy over 40% of the investigated forest. Mass occurrence and very low survivorship of seedlings under *A. distentifolium* are decisive for the rapid reduction of the spruce cohort after the year of mast seeding. The high mortality of spruce seedlings is likely related to the small amount of light reaching the forest floor under the fern. Small seedlings can also be buried under abundant fern litter.

Other ferns, *Dennstaedtia punctilobula* and *Osmunda claytoniana*, were found to have a similarly negative effect on tree regeneration in mixed forests of eastern North America (George and Bazzaz 1999a, b, Peterson and Pickett 2000, Cretaz et al. 2002). The mechanism of fern interference varied between tree species, but the influence was negative in all cases. One of the most important was the light reduction under fern fronds resulting in low growth rate of seedlings and their low survival.

The role of mosses in spruce regeneration is relatively small in the studied forest. Only *Polytrichum formosum* forms dense turfs, covering as little as 2.5% of the area. The fate of seedlings within this orthotrophic moss was similar to that under *A. distentifolium*. Hörnberg et al. (1997) obtained similar results in boreal spruce forest with *Polytrichum commune*. The negative effect of *Polytrichum* turfs on spruce regeneration is probably related to their height and to the thick layer of dead shoots, which separates small seedlings from the mineral bed. According to Hörnberg et al. (1997) the large annual height increment of mosses can be another reason for seedling decline.

The Role of Windthrow Mounds and Coarse Woody Debris in Spruce Regeneration

Pioneers and conifers are species often recruited on exposed mineral material of windthrow mounds and deadwood (Putz 1983, Nakashizuka 1989, Lertzman 1992, Little et al. 1994, Yamamoto 1995). The regeneration of Norway spruce, with its preference for logs and stumps, is a typical example of this. It has been observed in a wide spectrum of forests including boreal zone (Hytteborn and Packham 1987, Jonsson 1990, Liu and Hytteborn 1991, Hofgaard 1993a, Kuuluvainen 1994, Hörnberg et al. 1995, 1997, Kuuluvainen and Juntunen 1998) and European mountain forests (Mayer et al. 1972, Reif and Przybilla 1995, Filová 1998). It can be concluded that Norway spruce is one of the species which regeneration is most dependent on dead wood and mounds. Probably only *Tsuga heterophylla* – a species repeatedly investigated in terms of preferred microsites – is more dependent on logs than *Picea abies* (Franklin and Hemstrom 1981, Deal et al. 1991, Lertzman 1992).

The density of germinants on logs and windthrow mounds was considerably lower than on soil. These differences were probably due to unequal seed deposition. Mounds, stumps and logs are elevated above the surroundings, and this can make it easier for seeds to blow off their surfaces (Harmon 1989). Interestingly, on soil, where the greatest density of germinants was noted, their mortality was also the highest. As a result, mounds and logs covering a small area of the forest floor supported a considerable number of seedlings and most saplings. These microsites can be considered safe for recruitment of young spruces. The high growth rate of saplings on mounds and logs strengthens this conclusion. Nevertheless, the strong positive relationship between spruce regeneration and elevated structures is not only the result of the favorable conditions they provide. This pattern also stems from the inferior position of small seedlings on soil, where they have to penetrate thick litter and moss turfs with their roots and are exposed to competitive pressure from herbaceous plants and dwarf shrubs.

The area of windthrow mounds is small in the subalpine forest of Babia Góra, and their effect on spruce regeneration is marginal. However, even this small effect is higher than that noted in boreal forests of Scandinavia (Hytteborn and Packham 1987, Hofgaard 1993a), and it is similar to results reported by Ulanova (2000). It should be pointed out that the percentage of saplings on mounds could be underestimated, as a number of saplings classified as growing on soil may have originally germinated on mounds that were leveled completely.

Spruce regeneration on the exposed mineral ground of mounds is initiated immediately after tree death, and the establishment of spruces lasts a very long time; even windthrows created 40–100 years ago (decomposition classes VI and VII – comp. Holeksa 2001, Holeksa et al. 2008) support many saplings about 50 cm high. Neither the height nor the density of saplings increase with ageing of mounds. These two facts and the quick growth of saplings on mounds suggest that sapling turnover on this microsite is more intensive than on the others; older individuals decay and are replaced by fast-growing and younger ones. The high mortality of saplings on mounds presumably is caused by changes in mound topography due to root decay (Beatty and Stone 1986) and intensive erosion of elevated ground in a harsh climate with high precipitation.

Spruce colonization of logs is different. Immediately after a tree is downed its trunk is completely inaccessible to all vascular plants. There is only small amount of mineral compounds in a fresh log, and it is liberated slowly through the processes of wood decay (Sollins et al. 1987). The smooth surface of a fresh log allows seeds to be blown or washed off, and the hard wood makes it very difficult for vascular plant root systems to develop. These unfavorable conditions for seed germination and seedling growth on logs last about 20 years in the subalpine forest of Babia Góra. This is considerably less than in boreal forest in northern Sweden, where its remnants are accessible for spruce regeneration only 50 years after tree death (Hofgaard 1993b). In Babia Góra, seedlings were observed occasionally even on logs in class I of decomposition, but only on basal parts infected with *Heterobasidion annosum*. On such trunks, seedlings settled exclusively in deep cracks created by the fall of the tree.

Abundant spruce regeneration is launched when a log reaches decomposition class III, at least 20 years after the death of the tree (Holeksa 2001, Zielonka 2006b, Holeksa et al. 2008). Shallow furrows appear, and a thin layer of epixilic mosses develops on the log surface after that time. These changes favor retention of small seeds of spruce and diaspores of other plants on the logs, and also accelerate the accumulation of litter and its decomposition products (Harmon 1987).

Spruces higher than 30 cm were not found on most logs younger than 40 years (decomposition classes I-III). Sparse saplings were situated only on basal fragments of these logs. Logs in class III still present obstacles to seedling growth; survivorship of seedlings on them was lower than on more decayed debris. On the other hand, establishment and survivorship of cohort '93 on logs in class III and older did not differ. Both results suggest that an outer layer of decomposed soft wood as thin as 1 cm is enough for germinants to develop but is too thin for older, 5-10-year-old seedlings.

Approximately 30 years after tree death, the log enters class IV of decomposition (Holeksa 2001, Zielonka 2006b, Holeksa et al. 2008) and offers suitable conditions for seedling establishment and further development. After that time the logs are covered with crevices about 0.5 cm deep and the soft outer layer of wood is about 3 cm thick. They are also partly in contact with the soil. The number of saplings increases considerably on logs in classes V and VI, that is, at least 40 years after tree death, when the soft layer of wood is about 5 cm thick and the log is in contact with the ground along its whole length. Still later, 60-80 years after tree death, the rate of seedling sapling recruitment decreases; this is manifested in lower seedling density and changes in the sapling height distribution. On logs in classes III–VI, most saplings are <70 cm, while on older deadwood such small ones are less numerous than taller ones. Franklin and Hemstrom (1981) and Nakamura (1987) also reported that regeneration is limited on heavily decomposed debris. Regeneration of trees is worse on old logs, usually due to the abundant occurrence of epixilic plants, mainly mosses, which form dense and thick carpets (Nakamura 1992, Harmon 1989, Harmon and Franklin 1989). Old logs in the Babia Góra forest are also covered with dense and high turfs of *Polytrichum formosum*.

The number of spruce seedlings and saplings increased along with mound and log size. There was a negligible number of saplings on windthrow mounds of small (<20 cm) trees; soil disturbance resulting from the fall of such thin trees is of nearly no importance for spruce recruitment. The same holds for deadwood of diameter <20 cm, which is abundant in the studied forest but is not significant in spruce regeneration. Very small mounds probably are covered quickly with field-layer plants spreading vegetatively. Small mounds and thin logs are not elevated above the soil enough to protect the tiny seedlings against competition for light from the surrounding vegetation. Similar findings have been reported from other forests (Webb 1988, Takahashi 1994, Reif and Przybilla 1995). The most important microsites for spruce regeneration are those created by the death of trees 30-50 cm in diameter. Mounds and logs resulting from the fall of still thicker trees are much more suitable, but they are rare in the forest. The preferences of spruce for thick logs suggests that regeneration of deadwood-dependent species cannot be initiated at every stage of forest development. Disturbances in young stands resulting in the deposition of many thin logs may not provide an appropriate substrate for regeneration.

Canopy Gaps as a Factor Driving Spruce Regeneration

Norway spruce is a moderately shade-tolerant tree (Ellenberg 1986). However, its light requirements vary according to climatic conditions, and are particularly high in subalpine localities (Pisek and Winkler 1959). This suggestion is supported by the results from the Babia Góra forest, where spruce regeneration highly depends on canopy gaps. Not all gaps provide favorable conditions for young spruces. Saplings occur more densely than under the canopy only in gaps of at least 200 m^2. A similar minimum size of gaps suitable for regeneration is indicated by the rate of sapling growth. Again, only in gaps >200 m^2 was it higher than under the canopy. These results clearly show that the death of an individual spruce is insufficient for development of the next tree generation in this subalpine forest. Moreover, sapling density high enough for development of a closed tree layer is observed only after stand breakdown over an area of >1000 m^2. Large gaps, however, are usually older due to their protracted and progressive enlargement (Holeksa and Cybulski 2001). Possibly the higher sapling density in them is the result of not only better insolation but also longer recruitment of young spruces. Most large gaps have deadwood representing a wide spectrum of decomposition classes (Holeksa and Cybulski 2001). Such variability of logs situated within an individual gap enables establishment of uneven-aged spruce groups on a local scale. Gap expansion has been found to be a common phenomenon in other forests (Lertzman and Krebs 1991, Kubota 2000).

Gaps enhance light availability on a much larger area than their projected outline on the forest floor (Canham 1988b), and canopy edges are suitable for advanced regeneration of some trees (Kubota 2000). Better spruce regeneration at canopy edges than in the forest interior was also found in the subalpine forest, but this response was considerable only around gaps larger than 800 m^2.

The relationship between spruce regeneration and gap size was quite different in Scandinavian boreal forests. In central Sweden, where Norway spruce is a moderately shade-tolerant species (Leemans and Prentice 1987), saplings grow mainly in small gaps created by the death of individual trees (Leemans 1990, 1991, Liu and Hytteborn 1991, Hörnberg et al. 1995), and many of them grow under closed stands (Hytteborn et al. 1987). Hytteborn and Packham (1987) and Lundqvist (1991) reported that most seedlings grew under tree crowns, and Dai (1996) even claimed that poor illumination favors germination and seedling development in such boreal forests.

The pattern found in boreal forests was confirmed in the subalpine forest only for seedlings on soil. These small individuals grow in high density in the understory in patches dominated by *V. myrtillus*. Their recruitment in gaps is reduced by dense cover of *A. distentifolium* and *Calamagrostis villosa*. The reverse is true for seedlings growing on mounds, stumps and logs. These microsites are created simultaneously with gaps and are usually situated within them. Logs cover a larger area and are thicker, that is, more suitable for regeneration, inside gaps than outside them (Holeksa and Cybulski 2001). Windthrow mounds occupy gap centers, and young spruces growing on them take advantage of the high light level reaching the forest floor. This is reflected in the higher growth rate of spruces on this than on the other types of microsites.

This study demonstrates that spruce recruitment in subalpine spruce forest is affected by canopy gaps, the size and age of logs and windthrow mounds, as well as the field-layer vegetation. Gaps by themselves cannot promote spruce regeneration because they are accompanied by unfavorable changes in the field-layer vegetation. The large and suppressed seedling bank existing on the soil prior to a disturbance is probably of very little importance for establishment of a new spruce generation, because these tiny individuals are extinguished in gaps under *A. distentifolium*. Soil disturbance and log deposition in gaps are thus indispensable for spruce recruitment. Without mounds and downed wood, regeneration rather would be curtailed after gap creation. However, slow decomposition of deadwood in disturbed areas prolongs the process of spruce recruitment. This microsite is accessible to young spruces only some decades after tree death. The debris of large trees provides favorable conditions for regeneration. This implies that abundant spruce recruitment on deadwood is possible after the breakdown of mature and old-growth stands. Disturbances in pole-sized stands may not lead to sufficient regeneration, as thin logs are hidden under high fronds of *Athyrium distentifolium*.

Acknowledgments

This study was funded by the Polish State Committee for Scientific Research (grant no. N304 362938).

REFERENCES

Adamczyk, B. 1989. Characteristics of soils of selected areas in the upper spruce forest zone of Mt. Babia Góra. Pages 70-77 *in* Š. Korpel', editor. Stav, vývoj, produkcné schopnosti a funkcné vyuzivanie lesov v oblasti Babej Hory a Pilska. Vysoká škola lesnícka a drevárska, Zvolen (in Polish).

Barkman, J. J. 1973. Synusial approaches to classification. Pages 435-491 *in* R. H. Whittaker, editor. Classification of plant communities. Junk, The Hague.

Beatty, S. W., Stone, E. L. 1986. The variety of soil microsites created by tree falls. *Canadian Journal of Forest Research* 16: 539-548.

Bernier, N., Ponge, J. F. 1994. Humus form dynamics during the sylvogenetic cycle in a mountain spruce forest. *Soil Biology and Biochemistry* 26: 183-220.

Boone, R. D., Sollins, P., Cromack Jr., K. 1988. Stand and soil changes along a mountain hemlock death and regrowth sequence. *Ecology* 69: 714-722.

Brokaw, N. V. L. 1987. Gap-phase regeneration of three pioneer tree species in a tropical forest. *Journal of Ecology* 75: 9-19.

Burton, P. J., Bazzaz, F. A. 1995. Ecophysiological responses of tree seedlings invading different patches of old-field vegetation. *Journal of Ecology* 83: 99-112.

Canham, C. D. 1988a. An index for understory light levels in and around canopy gaps. *Ecology* 69: 1634-1638.

Canham, C. D. 1988b. Growth and canopy architecture of shade-tolerant trees: response to canopy gaps. *Ecology* 69: 786-795.

Cretaz, de la, A. L., Kelty M. J. 2002. Development of tree regeneration in fern-dominated understories after reduction of deer browsing. *Restoration Ecology* 10: 416-426.

Dai, X. 1996. Influence of light conditions in canopy gaps on forest regeneration: a new gap light index and its application in a boral forest in east-central Sweden. *Forest Ecology and Management* 84: 187-197.

Deal, R. L., Oliver, C. D., Bormann, B. T. 1991. Reconstruction of mixed hemlock-spruce stands in coastal southeast Alaska. *Canadian Journal of Forest Research* 21: 643-654.

Denslow, J. S. 1987. Tropical rainforest gaps and tree species diversity. *Annual Review of Ecology and Systematics* 18: 431-451.

Diggle, P. J. 1983. Statistical analysis of spatial point patterns. Academic Press, London.

Ellenberg, H. 1986. Vegetation Mitteleuropas mit den Alpen. Verlag Eugen Ulmer, Stuttgart.
Fil'ová I. 1998. Regeneration processes of a natural forest in the National Reserve Babia hora. Pages 34-39 *in* M. Saniga and P. Jaloviar, editors. Stav, vývoj, produkcné schopnosti a využivanie lesov v oblasti Babej hory a Pilska. Lesnicka fakulta Technickej university, Zvolen (in Slovak with English summary).
Franklin, J. F., Hemstrom, M. A. 1981. Aspects of succession in the coniferous forests of the Pacific Northwest. Pages 212-229 *in* D. C. Darrell, H. H. Shugart, and D. B. Botkin, editors. Forest succession: concepts and application. Springer-Verlag New York.
George, L. O., Bazzaz, F. A. 1999a. The fern understory as an ecological filter: emergence and establishment of canopy-tree seedlings. *Ecology* 80: 833-845.
George, L. O., Bazzaz, F. A. 1999b. The fern understory as an ecological filter: growth and survival of canopy-tree seedlings. *Ecology* 80: 846-856.
Goldblum, D. 1997. The effects of treefall gaps on understory vegetation in New York State. *Journal of Vegetation Science* 8: 125–132.
Haase P. 1995. Spatial pattern analysis in ecology based on Ripley's K-function: Introduction and methods of edge correction. *Journal of Vegetation Science* 6: 575-582.
Harmon, M. E. 1987. The influence of litter and humus accumulations and canopy openness on *Picea sitchensis* (Bong.) Carr. and *Tsuga heterophylla* (Raf.) Sarg. seedlings growing on logs. *Canadian Journal of Forest Research* 17: 1475-1479.
Harmon, M. E. 1989. Retention of needles and seeds on logs in *Picea sitchensis - Tsuga heterophylla* forests of coastal Oregon and Washington. *Canadian Journal of Botany* 67: 1833-1837.
Harmon, M. E., Franklin, J. F. 1989. Tree seedlings on logs in *Picea-Tsuga* forests of Oregon and Washington. *Ecology* 70: 48-59.
Hofgaard, A. 1993a. Structure and regeneration patterns in a virgin *Picea abies* forest in northern Sweden. *Journal of Vegetation Science* 4: 601-608.
Hofgaard, A. 1993b. 50 years of change in a Swedish boreal old-growth *Picea abies* forest. *Journal of Vegetation Science* 4: 773-782.
Holeksa, J. 2001. Coarse woody debris in a Carpathian subalpine spruce forest. *Forstwissenschaftliches Centralblatt* 120: 256-270.

Holeksa, J. 2003. Relationship between field-layer vegetation and canopy openings in a Carpathian subalpine spruce forest. *Plant Ecology* 168: 57-67.
Holeksa, J., Cybulski, M. 2001. Canopy gaps in a Carpathian subalpine spruce forest. *Forstwissenschaftliches Centralblatt* 120: 331-348.
Holeksa, J., Parusel, J. B. 1989. Snow cover in the forest zones of the Babia Góra massif (West Carpathians). *Acta Biologica Montana* 9: 341-352.
Holeksa J, Zielonka T, Zywiec M. 2008. Modeling the decay of coarse woody debris in a subalpine Norway spruce forest of the West Carpathians, Poland. *Canadian Journal of Forest Research* 38: 415-428.
Hörnberg, G., Ohlson, M., Zackrisson, O. 1995. Stand dynamics, regeneration patterns and long-term continuity in boreal old-growth *Picea abies* swamp-forests. *Journal of Vegetation Science* 6: 291-298.
Hörnberg, G., Ohlson, M., Zackrisson, O. 1997. Influence of bryophytes and microrelief conditions on *Picea abies* seed regeneration patterns in boreal old-growth swamp forests. *Canadian Journal of Forest Research* 27: 1015-1023.
Hytteborn, H., Packham, J. R. 1987. Decay rate of *Picea abies* logs and the storm gap theory: a re-examination of Sernander plot III, Fiby Urskog, central Sweden. *Arboricultural Journal* 11: 299-311.
Hytteborn, H., Packham, J. R., Verwijst, T. 1987. Tree population dynamics, stand structure and species composition in the montane virgin forest of Vallibäcken, northern Sweden. *Vegetatio* 72: 3-19.
Jaederlund, A., Zackrisson, O., Nilsson, M.-C. 1996. Effects of bilberry (*Vaccinium myrtillus* L.) litter on seed germination and early seedling growth of four boreal tree species. *Journal of Chemical Ecology* 22: 973-986.
Jonsson, B. G. 1990. Treefall disturbance – a factor structuring vegetation in boreal spruce forests. Pages 89-98 *in* F. Krahulec, A. D. Q. Agnew, S. Agnew, and J. H. Willems, editors. Spatial processes in plant communities. SPB Publisher, The Hague.
Kubota, Y. 2000. Spatial dynamics of regeneration in a conifer/broad-leaved forest in northern Japan. *Journal of Vegetation Science* 11: 633-640.
Kuuluvainen, T. 1994. Gap disturbance, ground microtopography, and the regeneration dynamics of boreal coniferous forests in Finland: a review. *Annales Zoologici Fennici* 31: 35-51.
Kuuluvainen T., Juntunen, P. 1998. Seedling establishment in relation to microhabitat variation in a windthrow gap in a boreal Pinus sylvestris forest. *Journal of Vegetation Science* 9: 551-562.

Leemans, R. 1990. Sapling establishment patterns in relation to light gaps in the canopy of two primeval pine-spruce forests in Sweden. Pages 111-120 *in* F. Krahulec, A. D. Q. Agnew, S. Agnew, and J. H. Willems, editors. Spatial processes in plant communities. SPB Academic, The Hague.

Leemans, R. 1991. Canopy gaps and establishment patterns of spruce (*Picea abies* (L.) Karst.) in two old-growth coniferous forests in central Sweden. *Vegetatio* 93: 157-165.

Leemans, R., Prentice, C. 1987. Description and simulation of tree-layer composition and size distributions in a primaeval *Picea-Pinus* forest. *Vegetatio* 69: 147-156.

Lertzman, K. P. 1992. Patterns of gap-phase replacement in a subalpine, old-growth forest. *Ecology* 73: 657-669.

Lertzman, K. P., Krebs, C. J. 1991. Gap-phase structure of a subalpine old-growth forest. *Canadian Journal of Forest Research* 21: 1730-1741.

Little, R. L., Peterson, D. L., Conquest, L. L. 1994. Regeneration of subalpine fir (*Abies lasiocarpa*) following fire: effects of climate and other factors. *Canadian Journal of Forest Research* 24: 934-944.

Liu, Q., Hytteborn, H. 1991. Gap structure, disturbance and regeneration in a primeval *Picea abies* forest. *Journal of Vegetation Science* 2: 391-402.

Lundqvist, L. 1991. Some notes on the regeneration of Norway spruce on six permanent plots managed with single-tree selection. *Forest Ecology and Management* 46: 49-57.

Mallik, A. U., Pellissier, F. 2000. Effects of *Vaccinium myrtillus* on spruce regeneration: testing the notion of coevolutionary significance of allelopathy. *Journal of Chemical Ecology* 26: 2197-2209.

Mayer, H., Schenker, S., Zukrigl, K. 1972. Der Urwaldrest Neuwald bei Lahnsattel. *Centralblat für das Gesamte Forstwesen* 89: 147-190.

Nakamura, T. 1987. Bryophyte and lichen succession on fallen logs and seedling establishment in Tsuga-Abies forests of central Japan. *Symposia Biologica Hungarica* 35: 485-495.

Nakamura, T. 1992. Effect of bryophytes on survival of conifer seedlings in subalpine forests of central Japan. *Ecological Research* 7: 155-162.

Nakashizuka, T. 1989. Role of uprooting in composition and dynamics of an old-growth forest in Japan. *Ecology* 70: 1273-1278.

Obrębska-Starklowa, B. 1963. The climate of Babia Góra. Pages 41-62 *in* W. Szafer, editor. The Babia Góra National Park. Polish Academy of Sciences, Kraków (in Polish with English summary).

Parusel, J. B., Holeksa, J. 1991. Relationship between the herb layer and the tree stand in the spruce forest of the upper montane belt in the West

Carpathians. *Phytocoenosis* (N.S.) 3, *Supplementum Cartographiae Geobotanicae* 2: 223-229.

Pellissier, F., Trosset, L. 1989. Obstacle allelopathique a la germination de graines d'epicea et a la croissance de trois champignons ectomycorhiziens de cette essence. *Acta Biologica Montana* 9: 153-160.

Peterson, C. J., Pickett, S. T. A. 1995. Forest reorganization: a case study in an old-growth forest catastrophic blowdown. *Ecology* 76: 763-774.

Peterson, C. J., Pickett, S. T. A. 2000. Patch type influences regeneration in a western Pennsylvania, USA, catastrophic windthrow. *Oikos* 90: 489-500.

Pickett, S. T., White, P. S. (editors) 1985. The ecology of natural disturbance and patch dynamics. Academic Press, Orlando.

Pisek, A., Winkler, E. 1959. Licht- und Temperaturabhängikeit der CO_2 – Assymilation von Fichte (*Picea excelsa* Link.) Zirbe (*Pinus cembra* L.) und Sonnenblume (*Helianthus annus* L.). *Planta* 53: 532-550.

Putz, F. E. 1983. Treefall pits and mounds, burried seeds, and the importance of soil disturbance to pioneer trees on Barro Colorado Island, Panama. *Ecology* 64: 1069-1074.

Rehak, J. 1957 The natural regeneration of spruce in the montane Spruce forest region. Prace vyzkumneho Ustavu lesniho hospodarstwi a myslivosti CSR 13: 33-66 (in Czech with English summary).

Reif, A., Przybilla, M. 1995. Zur Regeneration der Fichte (*Picea abies*) in den Hochlagen des Nationalparks Bayerischer Wald. *Hoppea* 56: 467-514.

Sollins, P., Cline, S. P., Verhoeven, T., Sachs, D., Spycher, G. 1987. Patterns of log decay in old-growth Douglas-fir forests. Canadian *Journal of Forest Research* 17: 1585-1595.

Sutherland, E. K., Hale, B. J., Hix, D. M. 2000. Defining species guilds in the Central Hardwood Forest, USA. *Plant Ecology* 147: 1-19.

Takahashi, K. 1994. Effect of size structure, forest floor type and disturbance regime on tree species composition in a coniferous forest in Japan. *Journal of Ecology* 82: 769-773.

Ulanova, N. G. 2000. The effects of windthrow on forests at different spatial scales: a review. *Forest Ecology and Management* 135: 155-167.

Wayne, P. M. and Bazzaz, F. A. 1993. Morning vs afternoon sun patches in experimental forest gaps: consequences of temporal incongruency of resources to birch regeneration. *Oecologia* 94: 235–243.

Webb, S. L. 1988. Windstorm damage and microsite colonization in two Minnesota forests. Canadian *Journal of Forest Research* 18: 1186-1195.

Yamamoto, S.-I. 1995. Gap characteristics and gap regeneration in subalpine old-growth coniferous forests, central Japan. *Ecological Research* 10: 31-39.

Zielonka, T., Niklasson M. 2001. Dynamics of dead wood and regeneration pattern in natural spruce forest in the Tatra Mountains, Poland. *Ecollogical Bulletins* 49: 159-163.

Zielonka, T. 2006a. When does dead wood turn into a substrate for spruce replacement? *Journal of Vegetation Science* 17: 739-746.

Zielonka, T. 2006b. Quantity and decay stages of coarse woody debris in old-growth subalpine spruce forests of the western Carpathians, Poland. *Canadian Journal of Forest Research* 36: 2614-2622.

In: Spruce
Editors: K. I. Nowak and H. F. Strybel

ISBN 978-1-61942-494-4
© 2012 Nova Science Publishers, Inc.

Chapter 2

UNCULTURED ARCHAEA IN SPRUCE RHIZOSPHERES AND MYCORRHIZAS

Malin Bomberg
Faculty of Biosciences, Department of Biological and Environmental Sciences, University of Helsinki, Finland

ABSTRACT

Archaea are microorganisms belonging to the third domain of cellular life on earth. Despite being unicellular microorganisms, they differ from the bacteria in many ways, and actually share an evolutionary history with the Eukaryotic branch of life. Since their discovery in extreme habitats, they have during the last two decades been found to inhabit almost all environments on Earth. Specific linages of archaea have been found to live forest soil ecosystems, and they have been found to be especially associated with boreal forest tree roots and mycorrhizospheres. These archaea belong to the so called Group I.1c of uncultured Thaumarchaeota (formerly included in the Phylum Crenarchaeota), and euryarchaeotal linages phylogenetically falling with the generally extremely halophilic Halobacteriales, as well as with methylotrophic and acetoclastic methanogens belonging to the Methanolobus and Methanosaeta, respectively.

Norway spruce is one of the most common forest trees in the Fennoscandian boreal forest. Only few studies have thus far been conducted on archaea inhabiting the rhizosphere and mycorrhizosphere of boreal forest trees, and even fewer have concentrated on Norway spruce.

However, it has been shown that the tree species has profound effect on the community composition of the archaea in the rhizosphere. When the tree roots are colonized by ectomycorrhizal fungi, the effect of the fungus dominates and the effect of the tree decreases. A spruce seedling grown in natural humus from a Scots pine stand gathered more detectable archaea in its roots than a Scots pine seedling, but when grown in natural humus from a Norway spruce stand the detectable number of archaea in the Norway spruce roots was much reduced. It has been shown that different tree species affect the soil they live in, and that Norway spruce has a tendency to acidify the soil, while for example silver birch increase the soil pH. Acidic spruce needle litter also bring recalcitrant organic matter and phenolics to the soil, while birch litter increase soil organic matter and nitrogen. An even greater difference between tree species can be detected in the composition of bacterial groups in the rhizospheres and mycorrhizospheres of different boreal forest tree species. The fact that archaea are generally found only in the roots and rhizosphere, and not in the soil uncolonized by mycorrhizal fungi or tree roots indicate a relationship between the archaea and the ectomycorrhizal fungi and the tree.

This chapter will concentrate on the archaea detected in the roots and mycorrhizosphere of boreal forest Norway spruce in comparison to other boreal tree species, but will also shortly touch on the subject of bacteria.

1. INTRODUCTION

Life on Earth is divided into three domains, the Eukaryota, the Bacteria and the Archaea (Woese *et al.*, 1990). The Archaea are, like the Bacteria, single celled micro-organisms. The Archaea (first called Archaebacteria) appeared to be a form of ancient micro-organisms found only in extreme environments, such as highly saline brines, hot springs and anaerobic habitats.

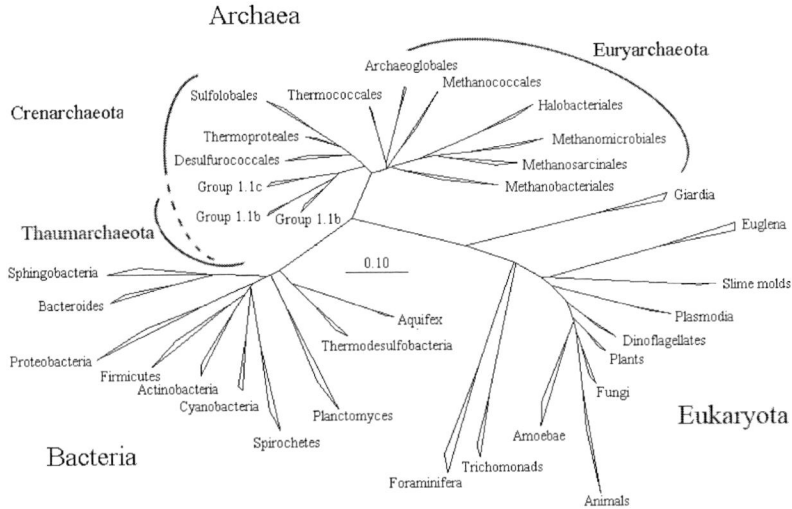

Figure 1. A unrooted phylogenetic tree displaying the three domains of life (modified from Bomberg, 2008). The dotted line describe groups formerly included in the Crenarchaeota, but which now are the Thaumarchaeota.

However, several studies on the phylogenetic relationships between Bacteria, Archaea and Eukarya have shown that the Archaea have shared an evolutionary history with the Eukaryota, which excluded the Bacteria, i.e. the Archaea appeared to be more closely related to the Eukaryota than to the Bacteria (Gogarten *et al.*, 1989; Iwabe *et al.*, 1989) (Figure 1).

During the last two decades, archaea have been found to be ubiquitiously present in all environments where they have been looked for and new specific archaeal linages have been found in different types of soil. Archaea have been shown to associate with the roots of agricultural crop plants as well as with natural plants in their native environments. Only few archaea of these soil lineages have yet been isolated in pure cultures and very little is known about their functions. However, certain groups of the soil archaea have been shown to especially associate with boreal forest tree roots and mycorrhizas, which indicates some kind of relationship between these groups of organisms.

1.1 The Boreal Forests

Coniferous forests grow in the northern parts of the Earth's northern hemisphere, the so called boreal regions. Boreal forests cover one third of the global forest areas and are thereby the largest forests in the world. They have a profound impact on global atmospheric CO_2 concentrations and the greenhouse effect. The boreal forests store about 60% of the global forest soil carbon and about 25% of the forest tree carbon (Dixon et al., 1994). The boreal forest soils are podzolised and typically consist of a top layer of organic humus (O) under which a thin, grayish-white eluvial (E) mineral soil horizon lies on top of the reddish-brown illuvial (B) mineral soil (e.g. Mokma et al., 2004, and references therein). In the boreal forests, the humus layer is typically thick because of the limited time of sufficiently high temperatures for decomposers to function. The highest respiration rates are found in the humus layer where particularly fungi and soil animals degrade organic litter, such as tree leaves and dead plant roots rich in lignin and cellulose (Lundström et al., 2000, and references therein). These degradation processes support the whole biological community of plants, animals and microbes. This microbial community further benefits the plants by forming diverse symbiotic relationships, such as mycorrhizas and symbiotic nitrogen-fixation (Tugel et al., 2000).

Boreal forests are typically dominated by coniferous trees. A typical boreal forest in the Fennoscandian region consist of Norway spruce, *Picea abies* (L) H. Karst. and Scots pine, *Pinus sylvestris* (L.). They have thick moss layers and are classified according to the under storey vegetation into *Oxalis-Myrtillum* type (OMT) which is the most humid type, to *Myrtillum* type (MT), *Vaccinum* type (VT), *Calluna* type (CT) and the driest *Calluna-Lichenes* type (CIT) forests (Cajander, 1926). Some deciduous tree species are also able to reproduce in the boreal areas. Of these the silver birch (*Betula pendula* Roth.) is the most common, although other birch species, such as paper birch (*B. pubescens* Ehrh.) can also be found.

Different tree species influence the physical, chemical and biological conditions of the soil in their vicinity by different means (Priha and Smolander, 1997). The pH of the soil in a coniferous forest is generally acidic and the typical Finnish dry pine forest soils exhibit a pH between 4 and 5. Conifers, especially spruce, may decrease the pH of the soil as low as 3.5, whereas birch has been shown to increase the soil pH up to 5.7 and enhance the cycling of nutrients (Mikola, 1985). The deciduous litter contains carbon, which is more easily leached, whereas coniferous litter is acidic and contains more stable, recalcitrant material such as waxes, phenolics and lignin (Harris and Safford, 1996; Prescott et al., 2004). Plants attract specific microbial communities to their rhizospheres by secreting different types of root exudates into the soil. Exudates, such as sugars, organic acids and amino acids attract among others, the ectomycorrhizal fungi living in the soil.

1.2 Ectomycorrhizal Fungi

The fine roots of boreal forest trees are generally colonized by ectomycorrhizal fungi, forming the symbiotic relationship called the ectomycorrhiza (ECM) between the plant host and the fungus (Figure 2). ECM are characteristic for most conifers, such as *Picea* and *Pinus*, and some Cupressaceae (e.g. *Juniperus*). They are also common in several dicot families such as the Betulacece, Fagaceae and Salicaceae. ECM have the ability, lacking in higher plants, to absorb and utilise ammonium directly from decaying organic material (Finlay *et al.*, 1992; van Hees *et al.*, 2006). They are mostly formed by basidiomyceteous fungi, but some Ascomycetes have also been reported to form ECM.

ECM root tips are primarily formed in the uppermost organic horizon and the interphase between the organic and mineral horizon of the podsolized soil of boreal forests (Smith and Read, 1997). Up to 90% of the ECM root tips are found in the O horizon (Jonsson *et al.*, 2000). More than 1000 species of fungi are estimated to form ECM in the roots of the few tree species in the Fennoscandian boreal forests (Knudsen and Hansen, 1991).

The mycorrhizal fungi have important impacts on the global carbon cycle. They have been shown to increase the rate of photosynthesis in the plants they colonize (Högberg *et al.*, 2008) thereby increasing the amounts of carbon the plant assimilates into its biomass and allocates into the soil. Not only is a considerable part of the plant derived carbon transported to the external mycelium, but carbon is also bound in the fungal biomass. The greatest part of the fungal carbon is in the form of membrane lipids, which are rapidly metabolized by other microbes when the fungal hyphae die. Another large proportion of fungal carbon is found in the chitin containing cell walls. These structures can be very resistant to microbial decomposers and be stored in the soil for years. The amount of fungal hyphae in the organic soil layer is considerable. It has been estimated that 125 to 200 kg of fungal hyphae are found per hectare of forest soil (Wallander *et al.*, 2001) and all ectomycorrhizal mycelia (including ectomycorrhizal mantle structures) were estimated to be as high as 700 to 900 kg ha^{-1}. The ECM fungal hyphae have been shown to conduct over 50% of the microbial respiration in boreal forest soil (Högberg *et al.*, 2001).

Figure 2. A Norway spruce seedling grown in a microcosm containing a thin layer of natural forest humus. The inserts show short toot tips colonized by different types of indigenous ectomycorrhizal fungi.

1.3. Mycorrhizosphere/Rhizosphere Concept

Mycorrhizospheres dominate nearly all natural soil ecosystems. However, pure, non-mycorrhizal rhizospheres can be found in nature. Non-mycorrhizal plant species are often the pioneers colonising land that has been disturbed by different means, such as forest fires and melting of glaciers (Allen, 1991). Terrestrial plant species which survive without mycorrhizal symbionts are usually found in relatively recent plant families such as Brassicaceae (e.g. Arabidopsis thaliana), Chenopodiaceae, Cyperaceae and Juncaceae (Brundrett, 1991).

The rhizosphere was described by Hiltner (1904) to be the soil adjacent to and influenced by plant roots. This compartment differs remarkably from the surrounding 'bulk' soil in pH, as well as concentration and quality of nutrients, micro-environments, moisture and oxygen levels (Nye, 1981; Wang and Zabowski, 1998). The mycorrhizosphere, on the other hand, was first described by Linderman (1988) as a term for the volume of soil influenced by the mycorrhizal fungus colonising the roots of a plant. Thus, the mycorrhizal fungi radically increase the amount of soil that can be exploited by a plant (Smith and Read, 1997).

Rhizospheres and mycorrhizospheres actively contribute to the heterogeneous structure of soil. Trees allocate 40%-70% of their photosynthetically assimilated carbon to the roots and 2-10% of this is further secreted as root exudates into the rhizosphere (reviewed by Grayston et al., 1997). Mycorrhizas have been shown to substantially increase the amounts of C the plant allocates to the roots and the plants often replace this carbon loss by increasing their level of photosynthesis. Low molecular weight, water soluble exudates, such as sugars, amino and organic acids, hormones and vitamins are diffused from the root tips into the surrounding soil (reviewed by Bertin et al., 2003) or are taken up by the mycorrhizal fungi directly from the plant root cells. The colonisation of the roots by mycorrhizal fungi alter both quality and quantity of exudates released into the soil. The fungus metabolizes the plant derived compounds into fungal metabolites, such as trehalose, mannitol and arabitol (Söderström et al., 1988) and they also produce lactic and oxalic acid. These compounds are secreted at the fast growing and metabolically active mycelial tips further modifying the soil habitats at the margins of the mycorrhizosphere and thereby attracting microorganisms to the rhizosphere (Finlay and Read, 1986).

1.4 Tree Rhizosphere Bacteria

Ectomycorrhizal fungi constitute the largest microbial biomass in forest soils. Although there are numerous species of mycorrhizal fungi able to colonize the roots of a tree species, the community is often dominated by only a few species of mycorrhizal fungi taking care of the everyday maintenance of the host (Heinonsalo *et al.*, 2007). Fungi often found in the tree mycorrhizospheres in the Fennoscandian boreal forest top soils are species of e.g. *Suillus*, *Paxillus*, *Tomentellopsis*, *Piloderma* (e.g. Kåren *et al.*, 1997; Timonen *et al.*, 1997).

Bacteria are highly abundant in soil and the rhizosphere. The highest bacterial abundances are found closest to the roots and are often relatively plant specific (Griffiths *et al.*, 1999; Duineveld *et al.*, 2001). In a study by Priha *et al.* (1997; 2001), the spruce rhizosphere soil in the organic soil layer contained twice the number of culturable bacteria (2.3×10^8 g^{-1}) than what was found for Scots pine, and one order of magnitude more than that of silver birch. The Norway spruce rhizosphere also had a higher number of both ammonia oxidizers and nitrate oxidizers than that of Scots pine or birch (Priha et al., 1999). From mycorrhizas of spruce seedlings in Slovenia, the most commonly found culturable bacterial species belonged to fluorescent *Pseudomonas* sp. and to *Bacillus* sp. (Geric et al., 2000). In boreal Norway spruce rhizosphere, pseudomonads were shown to constitute approximately 1% of the cultured bacteria (Priha et al., 2001). These bacterial species in addition to *Burkholderia* are also typically found in the mycorrhizosphere and rhizosphere of Scots pine (Timonen et al., 1998; Timonen and Hurek, 2006). In the montane forest of the Check Republic, Avidano et al. (2010) were able to show an even greater diversity of culturable bacteria, and reported members of the genera *Staphylococcaceae*, *Burkolderiaceae*, *Bacillaceae* and *Pseudomonaceae* to be the greatest groups of bacteria in Norway spruce mycorrhizal roots, with many other bacterial groups present as minor constituents of the community. The different habitats of the mycorrhizosphere, such as the mycorrhizas and external mycelial tips, have been shown to harbor bacterial communities differing in both numbers and activity (Timonen et al., 1998; Heinonsalo et al., 2001). In a microcosm experiment, Timonen et al. (1998) showed that the bacterial communities of mycorrhizospheres use different carbon sources than the ones in the forest humus uncolonised by mycorrhizal fungal hyphae. It was also shown that the bacterial communities inhabiting Scots pine-*Suillus bovinus* or -*Paxillus involutus* mycorrhizospheres favored different carbon sources. The *S. bovinus* communities especially favored mannitol, whereas the *P. involutus* communities preferred fructose.

Some bacteria have been shown to have great impact on the mycorrhization efficiency of fungi by producing substates such as IAA to attract fungi (reviewed by Garbaye, 1994; Frey-Klett et al., 2007). These bacteria were termed Mycorrhization Helper Bacteria, or MHBs. They represent diverse lineages and appear to show some specificity to types and species of mycorrhizal fungi. Bacteria in the mycorrhizosphere have been shown to promote germination of fungal spores (Xavier, 2003). Some bacteria have also been shown to increase the production of fungal mycelium and promote the branching of roots, which enhances the formation of mycorrhizas (Duponnois, 2006).

Non-symbiotic rhizosphere bacteria are known to promote plant growth, i.e. plant growth promoting rhizobacteria or PGPRs (reviewed in Zhuang et al., 2007). These bacteria may enhance plant growth by degrading organic compounds into suitable substrates for the plant. They reduce environmental stresses and protect the plants from pathogens. By production of gaseous substances, such as ethylene, these bacteria enhance root growth. They degrade environmental pollutants, such as PCB and PAHs. The Rhizobia are nitrogen fixing bacteria, which inside root nodules can live in symbiosis with leguminous plants. These microbes promote plant growth by fixing atmospheric nitrogen into NO_3^- and NH_4^+ which they release to the plant in return for constant and protected habitats with plant-provided carbon (Prell and Poole, 2006). They are also found freely in the rhizospheres of many non-nodulating plants (Maunuksela et al., 1999). The mycorrhizoshere and rhizosphere bacterial communities have also been reported to be involved in uptake and release of phosphorus from insoluble minerals (Leyval and Berthelin, 1993) and in degradation processes they are able to completely mineralize the compounds. The above mentioned activities of the bacteria in the rhizosphere are only a small part of the vast functional and taxonomical diversity found so far.

2. ARCHAEA

The archaea differ from the bacteria in many aspects (e.g. Cavicchioli, 2007; Garrett and Klenk, 2007). The major differences between the Archaea and Bacteria are the different lipids found in the membranes of the micro-organisms. While bacteria generally have a double membrane of phospholipids in which the fatty acids and glycerol moieties are ester-linked, the archaeal membranes are built up by branched-chain lipids containing hydrocarbon moieties bound to the glycerol by ether linkage. The archaeal information processing systems are also more reminiscent of the eukaryotal than the bacterial types. The bacterial cell walls contain peptidoglucan, which the archaeal cell walls lack. Instead, the archaea have diverse cell walls built of for example polysaccharides, glycoproteins or pseudopeptidoglucan. An important outcome of this is that antibiotics affecting Bacteria rarely harm the Archaea.

Today, the Archaea consist of two recognised phyla, the Euryarchaeota and the Crenarchaeota (Fig1). The Euryarchaeota contain all currently known methanogenic micro-organisms, the orders Methanopyrales, Methanococcales, Methanobacteriales, Methanomicrobiales and Methanosarcinales. The Methanopyrales appear to be rapidly evolving organisms and are phylogenetically situated far from the other methanogens (Brochier et al., 2004). The other methanogens are closely related to the extreme halophiles of the order Halobacteriales. Members of the last mentioned orders have recently been detected in moderate environments (e.g. Elshahed et al., 2004; Purdy et al., 2004; Bomberg and Timonen, 2009; Bomberg et al., 2011). Some taxa of the mainly thermophilic order Thermoplasmatales have also been detected in psycrophilic to mesophilic soil environments (Pesaro and Widmer, 2002). Members of the thermophilic orders Thermococcales and Archaeoglobales have so far only been found in extreme environments (e.g. Cavicchioli, 2007).

After the introduction of culture-independent molecular biological tools into microbial ecology, new types of Crenarchaeota have been found to be distributed ubiquitously in diverse moderate environments, such as soils and rhizospheres (e.g. Schleper *et al.*, 2005; Timonen and Bomberg, 2009). Crenarchaeal 16S rRNA genes were found both in aquatic environments and in agricultural (Bintrim *et al.*, 1997, Wisconsin; Großkopf *et al.*, 1998, Italy) and tropical forest soils (Borneman and Tripplett, 1997, Brazil). Studies on the soil of a boreal coniferous Norway spruce dominated forest in southern Finland revealed a novel, unique type of crenarchaeotal 16S rRNA genes not yet found anywhere else (Jurgens *et al.*, 1997; Jurgens and Saano, 1999).

Although members of the different euryarchaeotal orders have been found in moderate environments, the types of Crenarchaeota detected in these environments form their own monophyletic lineage affiliated with the Crenarchaeota, the Marine Group I of uncultured archaea (DeLong, 1998). It was discovered that the Crenarchaeota from the different moderate habitats differed significantly from each other in their 16S rRNA gene sequences and in 1998 DeLong described the phylogenetic grouping of these non-thermophilic Crenarchaeota. Group I of non-thermophilic Crenarchaeota contains three distinct clusters, groups I.1 (marine plankton, soil, sediments), I.2 (marine and lake sediments) and I.3 (lake sediments, palaeozols, anaerobic digesters). Group I.1 has since been divided into several subgroups. Group I.1a includes mainly planktonic archaea and group I.1b mostly archaea from soils and lake sediments. The boreal forest soil Crenarchaeota were so different from the other soil Crenarchaeota that they formed a separate group on their own, the I.1c group (Jurgens *et al.*, 1997).

Until recently, the Group I Crenarchaeota had no cultivated members. However, Quaiser *et al.* (2002) were able to isolate and sequence a large genome fragment of an uncultured I.1b crenarchaeote and Hallam *et al.* (2006b) succeeded in enriching the I.1a crenarchaeote, *Cenarchaeum symbiosum,* together with its sponge host and study its genome. Könneke *et al.* (2005) were the first to grow a crenarchaeote, *Nitrosopumilus maritimus,* in pure culture and this was phylogenetically placed within the Group I.1a lineage. Additionally, Tourna *et al.* (2011) managed to isolate and study a new ammonia oxidizing archaeon, *Nitrosophaera viennensis*, which was isolated from Austrian garden soil. This strain cluster phylogenetically to the Group

I.1b and was shown to be able to grow chemolithoautotrophically on ammonia or urea as energy source, but cultivations were shown to reach higher cell numbers when a carbon source in form of pyruvate was added. Interestingly, in co-culture with certain bacteria, the growth of the *N. viennensis* was also significantly increased. Lehtovirta-Morley *et al* (2011) enriched an autotrophic ammonia-oxidizing archaeon *Candidatus* '*Nitrosotalea devanaterra*' belonging to the Group I.1a associated Thaumarchaeota archaeota from upland pasture soil. This archaeon grew in laboratory culture only in the presence of a bacterial partner. Many of the ammonia oxidizing archaea have been shown to grow autotrophically by the fixation of CO_2 (Offre *et al.*, 2010). The growing number of whole-genome studies strengthen the suggestion by Brochier-Armanet *et al.* (2008) that this Group I Crenarchaeota actually forms a distinct new phylum, the Thaumarchaeota (Figure 1). In this chapter, the name Thaumarchaeota will henceforth be used.

2.1 Soil Archaea

Both boreal and temperate forest soils have been shown to emit methane (Kusel and Drake, 1994; Yavitt *et al.*, 1995; Kusel *et al.*, 1999; Fritze *et al.*, 1999; Sinha *et al.*, 2007; Bomberg *et al.*, 2010) and it has recently been shown that methanogenic archaea belonging to the genera Methanolobus and Methanosaeta colonize various parts of the rhizosphare and mycorrhizosphere of boreal forest trees (Bomberg *et al.*, 2010; 2011). Other types of archaea are continuously detected in both natural and agricultural soil. In water saturated soils, specific clusters of methanogenic Euryarchaeota have been found in rice field soils (e.g. Erkel *et al.*, 2005; Sakai *et al.*, 2007) and peatland soil (Brauer *et al.*, 2006). Non-extremophilic members of the Thermoplasmatales are often found in aquatic ecosystems (Jurgens *et al.*, 2000) but were recently also detected in the soil of a Swiss deciduous forest (Pesaro and Widmer, 2002). Members of the hyperhalophilic order Halobacteriales in boreal forest soil and tree mycorrhizospheres and roots, which are non-saline environments, have been reported in several studies (Jurgens *et al.*, 1996 - GeneBank submission; Bomberg and Timonen, 2007; 2009; Bomberg *et al.*, 2010; 2011).

In dry soils Thaumarchaeota are the predominant type of archaea. It has been estimated that the relative abundance of thaumarchaeotal 16S rRNA genes in agricultural and natural field soils is up to 1 to 2% of the total 16S rRNA gene pool (Buckley *et al.*, 1998; Sandaa *et al.*, 1999) and 0.3-0.5% in sandy soil (Ochsenreiter *et al.*, 2003). The dominating type of Thaumarchaeota in most aerated soils, both natural and agricultural, is Group I.1b (Nicol and Schleper, 2006). Group I.3 Thaumarchaeota have also recently been found in mature Austrian alpine grass land soil and Australian agricultural soil (Nicol *et al.*, 2005; Midgley *et al.*, 2007). The predominant group of Thaumarchaeota in the acidic (pH 3.5 - 5) coniferous boreal forests in southern Finland and in Scotland are the I.1c (Jurgens *et al.*, 1997; Jurgens and Saano, 1999; Bomberg *et al.*, 2003; Yrjälä *et al.*, 2004; Bomberg and Timonen, 2007; Nicol *et al.*, 2007), but also representatives belonging to different euryarchaeotal clades have been obtained, such as sequences belonging to methanogenic archaea and Halobacteriales (Bomberg and Timonen, 2007; 2009; Bomberg *et al.*, 2010; 2011).

The distribution of group I.1c Thaumarchaeota has been suggested to be affected by the pH of soil. In a few studies, both group I.1b and I.1c have been reported simultaneously, but I.1c Thaumarchaeota have with only one exception (Yrjälä *et al.*, 2004) not been found in soil with a pH above 5.1. In addition, this approaches the pH minimum in which I.1b Thaumarchaeota have usually been detected (Nicol *et al.*, 2003a; Oline *et al.*, 2006; Hansel *et al.*, 2008). The I.1c Thaumarchaeota are not specific to the boreal forest soils and have been detected in various mature and unmanaged grassland soils where the soil pH was below 5 (Nicol *et al.*, 2003a; Nicol *et al.*, 2003b; Ochsenreiter *et al.*, 2003; Nicol *et al.*, 2005; Hansel *et al.*, 2008). It is possible, however, that the pH is a secondary feature of the habitat, and that the appearance of I.1c Thaumarchaeota depends on other factors such as the presence of mycorrhizal fungi and their exudates.

2.2 Archaea in Roots and Rhizosphere Soil

Group I.1b Thaumarchaeota have been found to inhabit the root system of tomato plants grown in agricultural soil (Simon *et al.*, 2000). A few Group I.1a Thaumarchaeota were detected by archaea-specific PCR from maize roots (Chelius and Triplett, 2001), although this group is more commonly found in aquatic ecosystems. Both above mentioned studies were performed on agricultural soil from Wisconsin. Other studies have examined archaeal communities in the rhizospheres of plants growing in pristine soils (Sliwinski and Goodman, 2004; Nicol *et al.*, 2005) with the only type of archaea found belonged to Group I.1b Thaumarchaeota. In the study by Sliwinski and Goodman (2004), there were no differences in the populations of archaea inhabiting the roots or rhizosphere soil between plants of different genera growing in the same location. The same was seen in the study by Nicol et al. (2005), when they examined the rhizosphere soil of dominant plant species colonizing the different locations of a gradient of maturing soil in front of a receding glacier. Both studies concluded that there was a greater variability of the archaea in the soil between sites than in the rhizosphere of different plant genera within a site. Recent studies have shown that both the Thaumarchaeota of Group I.1a and I.1b carry genes which are involved in ammonia oxidation (Treuch et al., 2003; Auguet et al., 2008; Lehtovirta-Morley et al., 2011), and that the Group I.1a cluster are also able to live autotrophically by CO_2 fixation (Auguet et al., 2008; Lehtovirta-Morley et al., 2011). Leininger et al. (2006) have found that the ammonia oxidizing archaea predominate over bacteria in soil. However, in comparison to what is known about soil, plant and fungal associated bacteria, very little is known about the archaea found in these habitats.

Quantification of the number of microbial cells in boreal humus by epifluorescence microscopy is challenging because of the high amount of organic particles in the samples. Many substances are also autofluorescent, which additionally hampers the interpretation of the results. However, it has been estimated by PFLA analysis of humus of a Norway spruce stand in Norway that the number of bacterial cells g^{-1} fresh weight humus is between 6.4×10^9 to 7.9×10^{10} (Bach et al., 2008). The number of archaeal cells in the boreal forest humus and rhizosphere/mycorrhizosphere environments is probably very low, although no quantifications have yet been reported. Nevertheless, in a study by Fritze et al. (1999), where the archaea-specific lipid archaeol was search for in the humus of a coniferous forest, no archaeol was found in the humus sample. The detection limit, which had been determined empirically by adding pure cultured cells of *Halobacterium salinarum* to humus samples, was 10^8 cells g^{-1} dry weight of soil. This leads to the suggestion that the number of archaeal cells in boreal forest humus is less than 10^8 cells g^{-1} dry weight humus and the relative abundance of archaea in this habitat is less than 1% of the total prokaryotic community. A hypotetical estimation of archaeal cell numbers in the ectomycorrhizisphere and boreal forest soil microcosms has been made, suggesting the number of archaeal cells in 1 g of fresh weight soil to be below 1000 cells (Bomberg, 2008). In the non-mycorrhizal short roots, the number of archaeal cells per root tip would be around 10 cells, but in the mycorrhizas the archaeal cells have been shown to be enriched and to be in the range of 25 to 250 archaeal cells per mycorrhizal root tip. The spruce roots and mycorrhizas appear to have less archaea, as shown in Bomberg *et al.* (2011). When spruce seedlings were grown with the ectomycorrhizal fungus *Paxillus involutus* in humus obtained from a Scots pine dominated forest stand, archaea were readily detected in spruce mycorrhizas (Bomberg and Timonen, 2009). However, when spruce seedlings were grown in humus obtained from a spruce dominated stand the detection of archaea was noticebly lower (Bomberg *et al.*, 2011). These estimations are several magnitudes lower than the numbers of archaeal 16S rRNA genes detected in other studies. Kemnitz *et al.* (2007) calculated by quantitative PCR that the number of archaeal 16S rRNA genes in the upper layers of a temperate mixed deciduous forest in Germany to be as high as 0.5 to 3.9×10^8 g^{-1} dry soil. Sandaa *et al.* (1999) calculated by FISH that a Norwegian agricultural field had between $2.6-4.2 \times 10^7$ archaeal cells g^{-1} soil, corresponding to 1-2% of the microbial population. The same level of archaeal relative abundance has been reported in other agricultural and field soils as well (Buckley *et al.*, 1998; Ochsenreiter *et al.*, 2003). In the German deciduous forest soil, Kemnitz *et al.* (2007) showed that the archaea constituted a considerably greater part of the prokaryotic community (12-38%) than that detected in the agricultural field soil. No comparisons between thea:bacteria ratio in boreal forest soils have yet been reported.

The majority of the archaeal 16S rRNA genes detected in Norway spruce and other boreal forest tree mycorrhizospheres and humus have belonged to Group I.1c Thaumarchaeota. This group was also the most abundant type of archaea detected by Kemnitz *et al.* (2007) and constituted up to 85% of the archaeal community in the deciduous forest soil. However, the majority of archaea found in other types of soils, such as tropical forest soil, emerging glacier front soil, sandy grass fileld soils, agricultural soil, belonged to I.1b Thaumarchaeota (e.g. Buckley *et al.*, 1998; Ochsenreiter *et al.*, 2003; Nicol and Schleper, 2006). In a study by Burke *et al.* (2002) the effect of AM mycorrhiza on the abundance of rhizosphere microorganisms was tested with *Spartina patens* growing in a salt marsh. Archaeal cell numbers of a magnitude of 10^7 cells g^{-1} dry weight soil were estimated by FISH. It was also found that the abundance of AM fungi in the salt marsh soil had little effect on the number of archaeal cells. The type of archaea was not identified by Burke *et al.*, but compared to the study by Purdy *et al.* (2004) in a similar habitat, it is likely that the archaea detected in the hybridization experiments were euryarchaeotal.

Simon et al. (2000) used epifluorescent microscopy with FISH probes to estimate the number of archaeal cells on tomato roots. A 10-fold higher number of archaeal cells were detected on senescent roots compared to young roots. Although not as dramatic, bacteria were also more abundant on senescent rootlets compared to young roots. The population of I.1b Thaumarchaeota on the tomato roots was as high as 200 cells per rootlet. However, only the senescent tomato roots reached this number. Generally, the rootlets presented between three and 10 archaeal cells per rootlet. This is in agreement with the estimations for ectomycorrhizas and non-mycorrhizal roots of boreal forest trees (Bomberg, 2008).

In microcosm experiments, archaea have been found in all compartments of the tree rhizosphere, mycorrhizosphere and humus (Bomberg *et al*, 2003; 2010; 2011; Bomberg and Timonen, 2007; 2009). However, a clear difference in archaeal community composition and detection frequency between the different compartments has been observed. With only a few exceptions, the detection frequency and diversity of archaeal OTUs (RFLP groups or DGGE bands) was reported to be highest in the mycorrhizas and lowest in the non-mycorrhizal fine roots and humus devoid of fungal hyphae. In general, the average number of different archaeal phylotypes in the mycorrhizas and non-mycorrhizal roots varies between five and 10 when detected by DGGE. Sliwinski and Goodman (2004) reported a similar richness of Thaumarchaeota in different plant roots. However, a much lower number of different archaeal 16S rRNA genes have in general been detected in the bulk soil than in the rhizosphere and mycorrhizosphere (Bomberg and Timonen, 2007; 2009; Bomberg *et al.*, 2011; Sliwinski and Goodman, 2004). The non-mycorrhizal short roots of boreal forest trees have been reported to sustain a high diversity of bacteria (e.g. Timonen *et al.*, 1998; Timonen and Hurek, 2006). That is generally not the case with archaea. Nevertheless, the growing tips of the long roots have been shown to have as many different archaeal phylotypes as the mycorrhizas in Scots pine and silver birch, but not, however, in Norway spruce (Bomberg *et al.*, 2011).

2.3 Phylogeny of Archaea in Boreal Forest Norway Spruce Roots and Mycorrhizas

Almost all Thaumarchaeota found in the boreal forest soil and mycorrhizosphere so far belong to the Group I.1c although the archaea generally found in soils belong to the I.1b group (Figure 3). The found I.1c Thaumarchaeota divide on three major groups of which the majority cluster with the so called Finnish Forest Soil archaea group B or FFSB type Thaumarchaeota. This is the most commonly found type of archaea detected in the boreal forest soil, which were first found in clear-cut and burned Norway spruce dominated coniferous forest soil (Jurgens and Saano, 1997). A few of the sequences detected in this study also fell with the group FFSC which contains sequences originally found in humus from an undisturbed Norway spruce forest (Jurgens and Saano, 1999), and with the so-called I.1c associated group FFSB6, both of which appear to be much rarer than the FFSB group. Only little research has so far been performed on the I.1c Thaumarchaeota, because their global distribution is limited to mostly acidic forest or field soils, where their population densities usually are low. The thaumarchaeotal 16S rRNA gene sequences obtained from non-mycorrhizal fine roots of different boreal tree species fall into a separate cluster of the I.1c Thaumarchaeota than the ones from mycorrhizas (Bomberg and Timonen, 2009)(Fig 3). This further implies that the archaeal populations in the different rhizospheric and mycorrhizospheric habitats may react to different factors provided by either the tree species or the mycorrhizal fungus.

Two types of euryarchaeotal 16S rRNA gene sequences have been found in the boreal tree mycorrhizospheres. The most frequently detected euryarchaeotal genes show high similarity to the 16S rRNA gene sequences belonging to the genus *Halobacterium* and *Methanosaeta*. The sequences of putatively moderate halophilic Euryarchaeota retrieved from other moderate environments, Zodletone Spring (Oklahoma, USA) (Elshahed *et al.*, 2004) and open mud pans of Colne estuary marshes (Essex, UK) (Purdy *et al.*, 2004), do not phylogenetically cluster with the sequences from boreal forest tree mycorrhizospheres and roots. The sequences belonging to *Methanosaeta* found in the rhizospheres and mycorrhizospheres of boreal forest trees (Bomberg *et al.*, 2011) have also been detected in freshwater sediments and minerotrophic fens (Juottonen *et al.*, 2008). The other type of methanogen found belonged to the genus *Metanolobus* (Bomberg *et al.*, 2010; 2011). Both genera belong to the order Methanosarcinales, which also includes the genus *Methanosarcina*. This is one of the most versatile groups of methanogens and they are able to use many different substrates for methanogenesis (Smith and Ingram-Smith, 2007). Some of these methanogens have even been shown to be aerotolerant to some extent (Erkel *et al.*, 2006) and are often encountered in soils which may be exposed to oxygen, such as rice (Lueders and Friedrich, 2000) and cereal fields (Poplawski *et al.*, 2007).

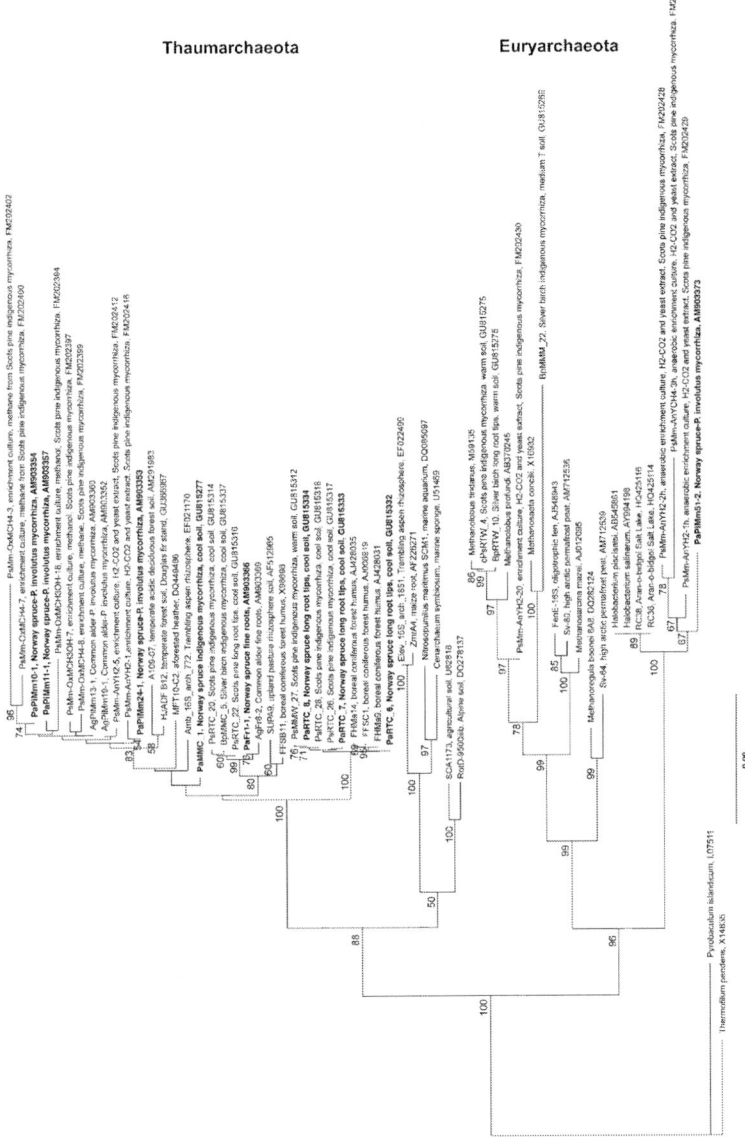

Figure 3. A maximum parsimony tree on the archaeal 16S rRNA gene sequences obtained from Norway spruce roots and mycorrhizas (bold) in relation to reference rhizosphere and soil archaea. Nodes with >50% bootstrap support are indicated. Bootstrap support values were calculated from 1000 random repeats.

2.4 Effect of Mycorrhizal Fungi on the Population of Archaea

The occurrence of group I.1c Thaumarchaeota has been linked to the presence of mycorrhizal fungi in the soil (Reviewed in Timonen and Bomberg, 2009). In the Austrian Alps, for instance, I.1c Thaumarchaeota only appeared in the soil when it was inhabited by mycotrophic plant species (Nicol *et al.*, 2005; Nicol *et al.*, 2006). Young soils without plant cover or inhabited by pioneering plants, which usually are non-mycorrhizal or only slightly mycotrophic, were devoid of I.1c Thaumarchaeota. Thaumarchaeota of the I.1b group have been shown to associate with plant roots and exist in soils with ECM and/or AM fungi (Sliwinski and Goodman, 2004). They have commonly been found also in non-vegetated soils, such as recently uncovered glacier front soil (Nicol *et al.*, 2005; Nicol *et al.*, 2006).

The species of ECM fungus colonizing the roots of forest trees have a great effect on the population of archaea harbored in the mycorrhizospheres. Without an ECM fungus, the Norway spruce short roots was shown to only supported a few archaea and the Scots pine supported no detectable archaeal populations (Bomberg and Timonen, 2009). However, when the Norway spruce rhizosphere was colonised by the ECM fungus *P. involutus*, the detection rate of archaea increased dramatically (Bomberg and Timonen, 2009). When the growing tips of the long roots of Norway spruce was examined, the situation was the opposite. More than three times higher detection rate of archaea was obtained from the growing root tips than from the mycorrhizas (Bomberg *et al.*, 2011).

The most commonly detected archaeal 16S rRNA gene sequence in the habitats provided by the different types of ectomycorrhizal fungi belonged to the genus *Halobacterium* (Bomberg and Timonen, 2007), except in the case of *P. involutus*, where most of the archaeal 16S rRNA gene sequences belonged to the I.1c Thaumarchaeota. The *Halobacterium*-like 16S rRNA genes have exclusively been detected in fungal samples, *i.e.* mycorrhizas, external mycelium and hyphal humus, but not in humus uncolonised by mycorrhizal fungal hyphae, or on non-mycorrhizal fine roots. Halobacterial populations in mycorrhizospheric compartments are frequently detected, which implies that these populations are stable and connected to the mycorrhizosphere. The thaumarchaeotal population dynamics have been shown to be much more sporadic and generally display mostly unique thaumarchaeotal OTUs between replicate samples. It appears as though the thaumarchaeotal diversity in hyphal and non-hyphal humus is relatively high although the detection frequency was low, indicating a rich population but very low cell numbers. Both the Thaum-

and Euryarchaeota detected in ectomycorrhizospheres appear to specifically inhabit mycorrhizal fungi containing habitats. This is in contrast to studies on the frequently detected soil inhabiting I.1b Thaumarchaeota. These Thaumarchaeota have been detected on the roots of plants, but do not appear to have specific preferences to certain plant species and they are present in the vicinity of both mycorrhizal and non-mycorrhizal plants (Sliwinski and Goodman, 2004; Nicol et al., 2005).

2.5 Specific effect of Tree Species on the Rhizospheric Archaeal Population

When the rhizospheres of different tree species are colonized by a certain species of ECM fungus, the archaeal communities between the mycorrhizospheres of the different tree species are surprisingly similar. This was shown with four different tree species, Norway spruce, Scots pine, paper birch and common alder, which all were colonized by *P. involutus* (Bomberg and Timonen, 2009). However, without the ECM, the different tree species had profound effect on the archaeal communities. The Scots pine fine roots did not harbor any detectable populations of archaea, Norway spruce and paper birch had some archaeal types, whereas the common alder roots showed a high diversity of archaeal types (Bomberg and Timonen, 2009). Alder is known to have nitrogen fixing microbial consortia in its rhizosphere, and to produce root nodules for the purpose, although no nodulation was detected in the mentioned experiments. Nevertheless, it has been shown, that alder is capable of capturing nodulating *Frankia* spp. in soils devoid of actinorhizal plants (Maunuksela et al., 1999). In a study by Priha et al. (1999), the Norway spruce had, however, a higher rate of ammonia and nitrate oxidation in the rhizosphere than both Scots pine and paper birch. Norway spruce and silver birch displayed only a low frequency and diversity of archaea in their non-mycorrhizal fine roots. However, although the Norway spruce and silver birch are known to have very different impacts on the soil in which they grow (Mikola, 1985), their archaeal populations were surprisingly similar in seedlings grown in the same soil (Bomberg and Timonen, 2009). The tree species appear to attract different numbers and different types of archaea. Although the Thaumarchaeota inhabiting tree roots cannot yet be determined to species level, the molecular characterization of the archaeal populations in the different tree roots may still reflect the same results as for the bacteria.

When the tree roots are colonised by *P. involutus*, the archaeal communities in all tested tree species became much more similar in composition.

All archaea found in the non-mycorrhizal roots and most of the ones from the *P. involutus* mycorrhizas have been I.1c Thaumarchaeota. However, sequences from non-mycorrhizal roots tend to cluster with different groups of I.1c Thaumarchaeota than those from mycorrhizas (Bomberg and Timonen, 2009). These results also agree with those obtained from bacterial studies. Timonen *et al.* (1998) and Timonen and Hurek (2006) have shown that the bacterial communities of non-mycorrhizal boreal forest tree roots are different from the bacterial population in the mycorrhizas. It is likely, that not only the impact of the tree, but also the population of rhizosphere inhabiting bacteria determine the number and composition of the archaeal populations. Archaea and bacteria are known to form syntrophic consortia in other environments (e.g. Orphan *et al.*, 2001b; Moissl *et al.*, 2002).

2.6 Anaerobic and Aerobic Archaea in the Boreal Forest Mycorrhizosphere

Very little is still known about the Group I.1c Thaumarchaeota found in different boreal forest soil environments, other than they appear to form higher community densities in the vicinity of mycorrhizas, mycorrhizal hyphae and boreal forest tree long root tips. They have recently been enriched in laboratory cultures both anaerobically and aerobically (Bomberg *et al.*, 2010). They appear to be heterotrophic, because they were anaerobically enriched in a medium containing yeast extract under an atmosphere of CO_2 and H_2 or methane and CO_2. When the carbon provided by the yeast extract was removed and the archaea were grown in a carbon-free mineral medium under the same gas atmosphere as mentioned above, only few I.1c thaumarcaotal types were obtained at extremely low frequency in the enrichments amended with methane and CO_2, whereas none was observed in the cultures amended only with CO_2 and H_2. They were also enriched aerobically in carbon-free mineral media to which methane and CO_2, or methanol was added to the enrichments, but again, no archaea were enriched in the aerobic carbon-free mineral medium to which only CO_2 and hydrogen had been added as substrates for growth. This strongly indicates that in contrast to other group I Thaumaerchaeota, especially the I.1a cluster and closely related archaea, the Group I.1c Thaumarchaeota are not autotrophic. Despite the fact that many Group I.1a and I.1b Thaumarchaeota have been shown to be ammonia oxidizers, no indications of this activity has been obtained by detection of ammonia monooxygenase genes in the studies of boreal forest rhizosphere and mycorrhizosphere archaea.

Nevertheless, four different pathways for CO_2 fixation have been identified in thermophilic Crenarchaeota (Hugler et al., 2003). These metabolic pathways also reflect the phylogenetic affiliation of the organisms. It is possible that the non-thermophilic I.1c Thaumarchaeota harbor yet another type of autotrophic CO_2 fixation. I.1a and I.1b Thaumarchaeota have been shown to grow aerobically (Simon et al., 2005; Könneke et al., 2005) and to oxidize ammonia (Könneke et al., 2005) or at least harbor the necessary genes for the ammonia monooxygenase (Treusch et al., 2005; Hallam et al., 2006a). Additionally, it has been suggested that the archaea probably play a greater role in the ammonia oxidation in soils than bacteria (Leininger et al., 2006). No such study has yet been reported for the I.1c Thaumarchaeota. It is possible that the I.1c Thaumarchaeota, which also are phylogenetically distant from the other Thaumarchaeota, do not participate in the oxidation of ammonia. The mycorrhizal fungi have a very efficient nitrogen metabolism and the plants also prefer ammonia, which might out-compete the archaeal ammonia oxidizers. This competition for ammonia may be one possible explanation for the lack of the commonly found I.1b Thaumarchaeota in acidic boreal forest soils. The possibility of ammonia oxidation cannot be excluded, but the detection of numerous I.1c thaumarchaeotal 16S rRNA genes in C-1 cycling enrichment cultures may indicate that the I.1c Thaumarchaeota may rather be involved in carbon cycling than in the oxidation of ammonia.

Halobacterial 16S rRNA genes have been identified in the non-saline anaerobic yeast extract enrichment cultures when CH_4 has either been added or produced. They have not been found in enrichment cultures on mineral media (Bomberg et al., 2010). This is in accordance with studies on extremely halophilic members of the Halobacteriales, which are known to be chemoorganotrophs (e.g. Gruber et al., 2004; Yang et al., 2006). Members of the genus *Halobacterium* are generally extremely halophilic. Moderately halophilic strains of typically halophilic genera have been isolated and detected by molecular methods from estuarine mud flats (Purdy et al., 2004) and a fresh water spring (Elshahed et al., 2004), but the isolated strains have still required at least 2.5% NaCl (Purdy et al., 2004). None of the 16S rRNA gene sequences of these new isolates clustered with the mycorrhizospheric halobacterial sequences in the phylogenetic analyses. Since the halobacteria were detected in anaerobic enrichments only when yeast extract and CH_4 were present, it may be suggested that the halobacteria have a role in the anaerobic cycling of methane, either by metabolizing it directly or by using derivatives produced by other microorganisms in the consortia. The halobacterial 16S rRNA sequences detected in the enrichment study by Bomberg et al. (2010) were all from anaerobic cultures. The Halobacteriales are generally aerobic microorganisms (Oren, 1999). However, the solubility of oxygen in hypersaline waters is low and the water often become anaerobic. Many of the Halobacteriales have been shown to have the capacity for anaerobic growth (Gruber et al., 2004; Muller and DasSarma, 2005; Yang et al., 2006), especially in the presence of certain amino acids. The medium used in Bomberg et al. (2010) was based on yeast extract, which contains amino acids, and may in the future be used for isolation of the non-halophilic Halobacterium-like Euryarchaeota residing in the boreal forest ectomycorrhizas.

Methanogenic archaea have been detected in the mycorrhizas and growing root tips of paper birch and Scots pine, but not from Norway spruce (Bomberg et al., 2010; 2011). Methanolobus sp. detected both in enrichments and directly from the roots and mycorrhizas of birch and pine is known to be methylotrophic, i.e. able to use methylated compounds, such as methanol and methylamines as carbon and energy source. Methansaeta sp., on the other hand, is one of the few methanogens able to utilize acetate. These substrates may be abundant in the rhizosphere, and in the rice roots, methanogens have been shown to produce CH_4 directly from root exudates (Lu and Conrad, 2005). CH_4 has been shown to be produced both from mycorrhizas and from boreal forest humus (Bomberg et al., 2010). However, the CH_4 production in enrichment cultures from mycorrhizas were shown to be over 10 times higher than from humus. In accordance to Lu and Conrad (2005) it is possible that also the mycorrhizospheric methanogens may be supported directly by exudates from the mycorrhizas. It was shown by Bomberg et al. (2010) that as soon as the enrichment medium was inoculated the CH_4 production started. This instantaneous high production of CH_4 supports the hypothesis that the mycorrhizospheric methanogens are aerotolerant and that the methanogenic community in the mycorrhizas is active and able to produce CH_4 as soon as favorable conditions occur. Methanogenic archaea show strong blue autofluorescence when excited with 420 nm UV light, due to their coenzyme F_{420}, and can be detected in unstained fresh samples by epifluorescence microscopy (Doddema and Vogels, 1978). When the enrichments of Bomberg et al. (2010) were examined in this way, comparably numerous methanogenic archaea with strong blue autofluorescence were detected. Most of the methanogens detected were encapsulated in a matrix together with nonfluorescent bacteria (as in Orphan *et al.*, 2001b). Such matrix may hamper DNA extraction and would explain the low detection level of sequences belonging to methanogenic archaea from boreal forest soil and rhizospheric environments. Members of the Methanosarcinales have shown a certain tolerance to oxygen and are one of the main groups found in rice field soil, where conditions shift between aerobic and anaerobic periods (Orphan *et al.*, 2001a; Orphan *et al.*, 2001b). Shifts in oxygen availability may also occur in the mycorrhizosphere due to microbial respiration (Li *et al.*, 1992).

CONCLUSION

Archaea have been shown to inhabit mycorrhizospheric habitats of boreal forest soil. They are rarely detected from bulk soil samples without any mycorrhizal fungal hyphae. However, whether saprotrophic fungi have the same effect has not yet been clarified. Both the type of ECM fungus and the species of the tree host has an effect on the population of archaea in the rhizosphere and mycorrhizosphere. The diversity of archaea and the success of their detection (apparently due to higher numbers of archaeal cells) is generally greater from mycorrhizal than non-mycorrhizal short roots. However, the growing tips of the lateral and vertical long roots have readily detectable archaeal populations. Most tested ECM fungi, with the exception of P. involutus, have been shown to support Euryarchaeota belonging to the genus Halobacterium, a group of archaea that is generally found in saline environments. P. involutus, on the other hand, have been show to enrich thaumarchaeotal types. The indigenous ECM picked up from natural humus have also showed that the archaeal communities contain methanogenic archaea. The colonization of tree roots by an ECM fungus clearly decreases the impact the tree species has on the archaeal community. Mycorrhizal I.1c Thaumarchaeota can be grown both anaerobically and aerobically at least on C-1 substrates, indicating that these archaea may play a role in both anaerobic and aerobic carbon cycling.

REFERENCES

Allen MF. (1991). *The ecology of mycorrhizae.* Cambridge University Press: Cambridge, NY. 184 pp.

Auguet J-C, Borrego CM, Bañeras L, Casamayor EO. (2008). Fingerprinting the genetic diversity of the biotin carboxylase gene (*acc*C) in aquatic ecosystems as a potential marker for studies of carbon dioxide assimilation in the dark. *Environ Microbiol* 10: 2527-2536.

Avidano L, Rinaldi M, Gindro R, Cudlín P, Martinotti MG, Fracchia L. (2010). Culturable bacterial populations associated with ectomycorrhizae of Norway spruce stands with different degrees of decline in the Czech Republic. *Can J Microbiol* 56: 52–64.

Bach LH, Frostegard Å, Ohlson M. (2008) Variation in soil microbial communities across a boreal spruce forest landscape. Canadian Journal of Forest Research 38: 1504-1516, 10.1139/X07-232

Bertin C, Yang X, Weston LA. (2003). The role of root exudates and allelochemicals in the rhizosphere. *Plant and soil* 256: 67-83.

Bintrim SB, Donohue TJ, Handelsman J, Roberts GP, Goodman RM. (1997). Molecular phylogeny of Archaea from soil. *Proc Natl Acad Sci U S A* 94: 277-282.

Bomberg M. (2008) Archaea in the mycorrhizosphere of boreal forest trees. Dissertation thesis, Faculty of Biosciences, University of Helsinki, ISBN 978-952-10-4726-8.

Bomberg M, Jurgens G, Saano A, Sen R, Timonen S. (2003). Nested PCR detection of archaea in defined compartments of pine mycorrhizospheres developed in boreal forest humus microcosms. *FEMS Microbiol Ecol* 43: 163-171.

Bomberg M, Timonen S. (2007). Distribution of Cren- and Euryarchaeota in Scots pine mycorrhizospheres and boreal forest humus. *Microb Ecol* 54: 406-416.

Bomberg M, Timonen S. (2009). Effect of tree species and mycorrhizal colonisation on the archaeal population of boreal forest rhizospheres. *Appl Environ Microbiol* 75: 308-315.

Bomberg M, Montonen L, Timonen S. (2010). Anaerobic Cren- and Euryarchaeota in boreal forest tree mycorrhiza. *Eur J Soil Biol* 46. 356-364.

Bomberg M, Münster U, Pumpanen J, Ilvesniemi H and Heinonsalo J (2011) Effect of temperature on the archaeal population in the rhizosphere and mycorrhizosphere of Scots pine, Norway spruce and silver birch. *Microb Ecol* 62: 205-17.

Borneman J and Triplett EW. (1997). Molecular microbial diversity in soils from eastern Amazonia: evidence for unusual microorganisms and

microbial population shifts associated with deforestation. *Appl Environ Microbiol* 63: 2647-2653.
Brauer SL, Cadillo-Quiroz H, Yashiro E, Yavitt JB, Zinder SH. (2006). Isolation of a novel acidiphilic methanogen from an acidic peat bog. *Nature* 442: 192-194.
Brochier C, Forterre P, Gribaldo S. (2004). Archaeal phylogeny based on proteins of the transcription and translation machineries: tackling the *Methanopyrus kandleri* paradox. *Genome Biol* 5: R17.
Brochier-Armanet C, Boussau B, Gribaldo S, Forterre P. (2008). Mesophilic crenarchaeota: proposal for a third archaeal phylum, the Thaumarchaeota. *Nat Rev Microbiol* 6: 245-252.
Brundrett M. (1991). Mycorrhizas in natural ecosystems. *Adv ecol res* 21: 171-313.
Buckley DH, Graber JR, Schmidt TM. (1998). Phylogenetic analysis of nonthermophilic members of the kingdom Crenarchaeota and their diversity and abundance in soils. *Appl Environ Microbiol* 64: 4333-4339.
Burke DJ, Hamerlynck EP, Hahn D. (2002). Effect of arbuscular mycorrhizae on soil microbial populations and associated plant performance of the salt marsh grass *Spartina patens*. *Plant and soil* 239: 141-154.
Cajander A K. (1926). The theory of forest types. *Acta For Fenn* 29: 85-108.
Cavicchioli R. (2007). *Archaea : molecular and cellular biology.* ASM Press: Washington, DC. 523 pp.
Chelius MK and Triplett EW. (2001). The Diversity of Archaea and Bacteria in association with the roots of *Zea mays* L. *Microb Ecol* 41: 252-263.
DeLong EF. (1998). Everything in moderation: archaea as 'non-extremophiles'. *Curr Opin Genet Dev* 8: 649-654.
Dixon RK, Solomon AM, Brown S, Houghton RA, Trexier MC, Wisniewski J. (1994). Carbon pools and flux of global forest ecosystems. *Science* 263: 185-190.
Doddema HJ and Vogels GD. (1978). Improved identification of methanogenic bacteria by fluorescence microscopy. *Appl Environ Microbiol* 36: 752-754.
Duineveld BM, Kowalchuk GA, Keijzer A, van Elsas JD, van Veen JA. (2001). Analysis of bacterial communities in the rhizosphere of *Chrysanthemum* via denaturing gradient gel electrophoresis of PCR-amplified 16S rRNA as well as DNA fragments coding for 16S rRNA. *Appl Environ Microbiol* 67: 172-178.
Duponnois R. (2006). The possible role of trehalose in the mycorrhiza helper bacterium effect. *Can Jour Bot* 84: 1005.

Elshahed MS, Najar FZ, Roe BA, Oren A, Dewers TA, Krumholz LR. (2004). Survey of archaeal diversity reveals an abundance of halophilic Archaea in a low-salt, sulfide- and sulfur-rich spring. *Appl Environ Microbiol* 70: 2230-2239.

Erkel C, Kube M, Reinhardt R, Liesack W. (2006). Genome of Rice Cluster I archaea-the key methane producers in the rice rhizosphere. *Science* 313: 370-372.

Erkel C, Kemnitz D, Kube M, Ricke P, Chin KJ, Dedysh S, Reinhardt R, Conrad R, LiesackW. (2005). Retrieval of first genome data for rice cluster I methanogens by a combination of cultivation and molecular techniques. *FEMS Microbiol Ecol* 53: 187-204.

Finlay RD and Read DJ. (1986). The structure and function of the vegetative mycelium of ectomycorrhizal plants. I. Translocation of 14C-labelled carbon between plants interconnected by a common mycelium. *New Phytol* 103: 143-156.

Finlay RD, Frostegard A, Sonnerfeldt A. (1992). Utilization of organic and inorganic nitrogen sources by ectomycorrhizal fungi in pure culture and in symbiosis with *Pinus contorta* Dougl. ex Loud. *New Phytol* 120: 105-115.

Frey-Klett P, Garbaye J, Tarkka M. (2007). The mycorrhiza helper bacteria revisited. *New Phytol* 176: 22-36.

Fritze H, Tikka P, Pennanen P, Saano A, Jurgens G, Nilsson M, Bergman I, Kitunen V. (1999). Detection of Archaeal Diether Lipid by Gas Chromatography from Humus and Peat. *Scand J For Res* 14: 545-551.

Garbaye J. (1994). Tansley Review No. 76 Helper bacteria: a new dimension to the mycorrhizal symbiosis. *New Phytol* 128: 197-210.

Garrett RA and Klenk H. (2007). *Archaea : evolution, physiology, and molecular biology.* Blackwell Pub: Malden, MA. 388 pp.

Geric B, Rupnic M, Kraigher H. (2000). Isolation and Identification of Mycorrhization Helper Bacteria in Norway spruce, *Picea abies* (L.) Karst. *Phyton (Austria) Special issue: "Root-soil interactions"* 40: 65-70.

Gogarten JP, Kibak H, Dittrich P, Taiz L, Bowman EJ, Bowman BJ, Manolson MF, Poole RJ, Date T, Oshima T, Konishi J, Denda K, Yoshida M. (1989). Evolution of the vacuolar H^+-ATPase: Implications for the Origin of Eukaryotes. *Proc Natl Acad Sci U S A* 86: 6661-6665.

Grayston SJ, Vaughan D, Jones D. (1997). Rhizosphere carbon flow in trees, in comparison with annual plants: the importance of root exudation and its impact on microbial activity and nutrient availability. *Appl Soil Ecol,* 5: 29-56.

Griffiths BS, Ritz K, Ebblewhite N, Dobson G. (1999). Soil microbial community structure: Effects of substrate loading rates. *Soil Biol Biochem,* 31: 145-153.

Großkopf R, Stubner S, Liesack W. (1998). Novel euryarchaeotal lineages detected on rice roots and in the anoxic bulk soil of flooded rice microcosms. *Appl Environ Microbiol* 64: 4983-4989.

Gruber C, Legat A, Pfaffenhuemer M, Radax C, Weidler G, Busse HJ, Stan-Lotter H. (2004). *Halobacterium noricense* sp. nov., an archaeal isolate from a bore core of an alpine Permian salt deposit, classification of *Halobacterium* sp. NRC-1 as a strain of *H. salinarum* and emended description of *H. salinarum*. *Extremophiles* 8: 431-439.

Hallam SJ, Mincer TJ, Schleper C, Preston CM, Roberts K, Richardson PM, DeLong EF. (2006a). Pathways of carbon assimilation and ammonia oxidation suggested by environmental genomic analyses of marine Crenarchaeota. *PLoS Biol* 4: e95.

Hallam SJ, Konstantinidis KT, Putnam N, Schleper C, Watanabe Y, Sugahara J, Preston C, de la Torre J, Richardson PM, DeLong EF. (2006b). Genomic analysis of the uncultivated marine crenarchaeote *Cenarchaeum symbiosum*. *Proc Natl Acad Sci U S A* 103: 18296-18301.

Hansel CM, Fendorf S, Jardine PM, Francis CA. (2008). Changes in Bacterial and Archaeal community structure and functional diversity along a geochemically variable soil profile. *Appl Environ Microbiol* 74: 1620-1633.

Heinonsalo J, Koskiahde I, Sen R. (2007). Scots pine bait seedling performance and root colonizing ectomycorrhizal fungal community dynamics before and during the 4 years after forest clear-cut logging seasons after forest clear-cut logging. *Can J For Res* 37: 415-429.

Heinonsalo J, Jørgensen KS, Sen R. (2001). Microcosm-based analyses of Scots pine seedling growth, ectomycorrhizal fungal community structure and bacterial carbon utilization profiles in boreal forest humus and underlying illuvial mineral horizons. *FEMS Microbiol Ecol* 36: 73-84.

Hiltner L. (1904). Über neuere Erfahrungen und Probleme auf dem Gebiete der Bodenbakteriologie unter besonderer Berücksichtigung der Gründüngung und Brache. *Arb Dtsch Lanwirt Ges* 98: 59-78.

Högberg P, Nordgren A, Buchmann N, Taylor AF, Ekblad A, Högberg MN, Nyberg G, Ottosson-Löfvenius M, Read DJ. (2001). Large-scale forest girdling shows that current photosynthesis drives soil respiration. *Nature* 411: 789-792.

Högberg P, Högberg MN, Gottlicher SG, Betson NR, Keel SG, Metcalfe DB, Campbell C, Schindlbacher A, Hurry V, Lundmark T, Linder S, Näsholm T. (2008). High temporal resolution tracing of photosynthate carbon from the tree canopy to forest soil microorganisms. *New Phytol* 177: 220-228.

Hugler M, Huber H, Stetter KO, Fuchs G. (2003). Autotrophic CO_2 fixation pathways in archaea (Crenarchaeota). *Arch Microbiol* 179: 160-173.

Iwabe N, Kuma K, Hasegawa M, Osawa S, Miyata T. (1989). Evolutionary relationship of Archaebacteria, Eubacteria, and Eukaryotes inferred from phylogenetic trees of duplicated genes. *Proc Natl Acad Sci U S A* 86: 9355-9359.

Juottonen H, Tuittila E-S, Juutinen S, Fritze H, Yrjälä K (2008). Seasonality of rDNA- and rRNA-derived archaeal communities and methanogenic potential in a boreal mire. *ISME Jour* 2:1157-1168.

Jurgens G and Saano A. (1999). Diversity of soil Archaea in boreal forest before, and after clear-cutting and prescribed burning. *FEMS Microbiol Ecol* 29: 205-213.

Jurgens G, Lindström K, Saano A. (1997). Novel group within the kingdom Crenarchaeota from boreal forest soil. *Appl Environ Microbiol* 63: 803-805.

Jurgens G, Glöckner F, Amann R, Saano A, Montonen L, Likolammi M, Münster U. (2000). Identification of novel Archaea in bacterioplankton of a boreal forest lake by phylogenetic analysis and fluorescent *in situ* hybridization(1). *FEMS Microbiol Ecol* 34: 45-56.

Kåren O, Högberg N, Dahlberg A, Jonsson L, Nylund J. (1997). Inter- and intraspecific variation in the ITS region of rDNA of ectomycorrhizal fungi in Fennoscandia as detected by endonuclease analysis. *New Phytol* 136: 313-325.

Kemnitz D, Kolb S, Conrad R. (2007). High abundance of Crenarchaeota in a temperate acidic forest soil. *FEMS Microbiol Ecol* 60: 442-448.

Knudsen H, Hansen L. (1991). New taxa and combinations in the *Agaricales*, *Boletales* and *Polyporales*. *Nord J Bot* 11: 477-481.

Könneke M, Bernhard AE, de la Torre JR, Walker CB, Waterbury JB, Stahl DA. (2005). Isolation of an autotrophic ammonia-oxidizing marine archaeon. *Nature* 437: 543-546.

Küsel K and Drake HL. (1994). Acetate Synthesis in Soil from a Bavarian Beech Forest. *Appl Environ Microbiol* 60: 1370-1373.

Küsel K, Wagner C, Drake HL. (1999). Enumeration and metabolic product profiles of the anaerobic microflora in the mineral soil and litter of a beech forest. *FEMS Microbiol Ecol* 29: 91-103.

Lehtovirta-Morley L, Stoecker K, Vilcinskas A, Prosser J, Nicol GW (2011) Cultivation of an obligate acidophilic ammonia oxidiser from a nitrifying acid soil. *Proc Nat Acad Sci USA*. *In press*.

Leininger S, Urich T, Schloter M, Schwark L, Qi J, Nicol GW, Prosser JI, Schuster SC, Schleper C. (2006). Archaea predominate among ammonia-oxidizing prokaryotes in soils. *Nature* 442: 806-809.

Leyval C, Berthelin J. (1993). Rhizodeposition and net release of soluble organic compounds by pine and beech seedlings inoculated with rhizobacteria and ectomycorrhizal fungi. *Biol Fertil Soils* 15: 259-267.

Li C, Massicote H, Moore LVH. (1992). Nitrogen-fixing *Bacillus* sp. associated with Douglas-fir tuberculate ectomycorrhizae. *Plant and soil* 140: 35-40.

Linderman RG. (1988). Mycorrhizal interactions with the rhizosphere microflora - the mycorrhizoshere effect. *Phytopathology* 78: 366-371.

Lu Y, Conrad R. (2005). *In situ* stable isotope probing of methanogenic archaea in the rice rhizosphere. *Science* 309: 1088-1090.

Lueders T, Friedrich M. (2000). Archaeal population dynamics during sequential reduction processes in rice field soil. *Appl Environ Microbiol* 66: 2732-2742.

Lundström US, van Breemen N, Bain D. (2000). The podzolization process. A review. *Geoderma,* 94: 91-107.

Maunuksela L, Zepp K, Koivula T, Zeyer J, Haahtela K, Hahn D. (1999). Analysis of Frankia populations in three soils devoid of actinorhizal plants. *FEMS Microbiology Ecology,* 28: 11-21.

Midgley DJ, Saleeba JA, Stewart MI, McGee PA. (2007). Novel soil lineages of Archaea are present in semi-arid soils of eastern Australia. *Can J Microbiol* 53: 129-138.

Mikola P. (1985). The effect of tree species on the biological properties of forest soil. *Naturvårdsverkets Rapport* 3017.

Moissl C, Rudolph C, Huber R. (2002). Natural communities of novel archaea and bacteria with a string-of-pearls-like morphology: molecular analysis of the bacterial partners. *Appl Environ Microbiol* 68: 933-937.

Mokma DL, Yli-Halla M, Lindqvist K. (2004). Podzol formation in sandy soils of Finland. *Geoderma* 120: 259-272.

Muller JA and DasSarma S. (2005). Genomic analysis of anaerobic respiration in the archaeon *Halobacterium* sp. strain NRC-1: dimethyl sulfoxide and trimethylamine N-oxide as terminal electron acceptors. *J Bacteriol* 187: 1659-1667.

Nicol GW, Schleper C. (2006). Ammonia-oxidising Crenarchaeota: important players in the nitrogen cycle? *Trends Microbiol* 14: 207-212.

Nicol GW, Glover LA, Prosser JI. (2003a). Spatial analysis of archaeal community structure in grassland soil. *Appl Environ Microbiol* 69: 7420-7429.

Nicol GW, Glover LA, Prosser JI. (2003b). The impact of grassland management on archaeal community structure in upland pasture rhizosphere soil. *Environ Microbiol* 5: 152-162.

Nicol GW, Campbell CD, Chapman SJ, Prosser JI. (2007). Afforestation of moorland leads to changes in crenarchaeal community structure. *FEMS Microbiol Ecol* 60: 51-59.

Nicol GW, Tscherko D, Embley TM, Prosser JI. (2005). Primary succession of soil Crenarchaeota across a receding glacier foreland. *Environ Microbiol* 7: 337-347.

Nicol GW, Tscherko D, Chang L, Hammesfahr U, Prosser JI. (2006). Crenarchaeal community assembly and microdiversity in developing soils at two sites associated with deglaciation. *Environ Microbiol* 8: 1382-1393.

Nye P E. (1981). Changes of pH across the rhizosphere induced by roots. *Plant and soil* 61: 7-26.

Ochsenreiter T, Selezi D, Quaiser A, Bonch-Osmolovskaya L, Schleper C. (2003). Diversity and abundance of Crenarchaeota in terrestrial habitats studied by 16S RNA surveys and real time PCR. *Environ Microbiol* 5: 787-797.

Offre PO, Nicol GW, Prosser JI. (2010) Autotrophic community profiling and quantification of putative autotrophic thuamarchaeal communities in environmental samples. *Environ Microbiol Rep* 3: 245-253.

Oline DK, Schmidt SK, Grant MC. (2006). Biogeography and landscape-scale diversity of the dominant Crenarchaeota of soil. *Microb Ecol* 52: 480-490.

Oren A. (1999). Bioenergetic aspects of halophilism. *Microbiol Mol Biol Rev* 63: 334-348.

Orphan VJ, House CH, Hinrichs KU, McKeegan KD, DeLong EF. (2001a). Methane-consuming archaea revealed by directly coupled isotopic and phylogenetic analysis. *Science* 293: 484-487.

Orphan VJ, Hinrichs KU, Ussler W, Paull CK, Taylor LT, Sylva SP *et al.* (2001b). Comparative analysis of methane-oxidizing archaea and sulfate-reducing bacteria in anoxic marine sediments. *Appl Environ Microbiol* 67: 1922-1934.

Pesaro M, Widmer F. (2002). Identification of novel Crenarchaeota and Euryarchaeota clusters associated with different depth layers of a forest soil. *FEMS Microbiol Ecol* 42: 89-98.

Poplawski AB, Mårtensson L, Wartiainen I, Rasmussen U. (2007). Archaeal diversity and community structure in a Swedish barley field: Specificity of the EK510R/(EURY498) 16S rDNA primer. *Journal of Microbiological Methods,* 69: 161-173.

Prell J, Poole P. (2006). Metabolic changes of rhizobia in legume nodules. *Trends in Microbiology,* 14: 161-168.

Prescott CE, Blevins LL, Staley C. (2004). Litter decomposition in British Columbia forests: Controlling factors and influences of forestry activities. *JEM* 5: 44--57.

Priha O, Smolander A. (1997). Microbial biomass and activity in soil and litter under *Pinus sylvestris*, *Picea abies* and *Betula pendula* at originally similar field afforestation sites. *Biol Fertil Soils* 24: 45-54.

Priha O, Smolander A. (1999). Nitrogen transformations in soil under *Pinus sylvestris*, *Picea abies* and *Betula pendula* at two forest sites. *Soil Biol Biochem* 31: 965–977.

Priha O, Grayston SJ, Hiukka R, Pennanen T, Smolander A. (2001). Microbial community structure and characteristics of the organic matter in soils under *Pinus sylvestris*, *Picea abies* and *Betula pendula* at two forest sites. *Biol Fertil Soils* 33:17–24.

Purdy KJ, Cresswell-Maynard TD, Nedwell DB, McGenity TJ, Grant WD, Timmis KN, Embley TM. (2004). Isolation of haloarchaea that grow at low salinities. *Environ Microbiol* 6: 591-595.

Quaiser A, Ochsenreiter T, Klenk HP, Kletzin A, Treusch AH, Meurer G, Eck J, Sensen CW, Schleper C. (2002). First insight into the genome of an uncultivated crenarchaeote from soil. *Environ Microbiol* 4: 603-611.

Sakai S, Imachi H, Sekiguchi Y, Ohashi A, Harada H, Kamagata Y. (2007). Isolation of key methanogens for global methane emission from rice paddy fields: a novel isolate affiliated with the clone cluster rice cluster I. *Appl Environ Microbiol* 73: 4326-4331.

Sandaa RA, Enger O, Torsvik V. (1999). Abundance and diversity of Archaea in heavy-metal-contaminated soils. *Appl Environ Microbiol* 65: 3293-3297.

Schleper C, Jurgens G, Jonuscheit M. (2005). Genomic studies of uncultivated archaea. *Nat Rev Microbiol* 3: 479-488.

Simon HM, Dodsworth JA, Goodman RM. (2000). Crenarchaeota colonize terrestrial plant roots. *Environ Microbiol* 2: 495-505.

Simon HM, Jahn CE, Bergerud LT, Sliwinski MK, Weimer PJ, Willis DK, Goodman RM. (2005). Cultivation of mesophilic soil crenarchaeotes in enrichment cultures from plant roots. *Appl Environ Microbiol* 71: 4751-4760.

Sinha V, Williams J, Crutzen P, Lelieveld J. (2007). Methane emissions from boreal and tropical forest ecosystems derived from in-situ measurements. *Atmos Chem Phys Discuss* 7: 14011-14039.

Sliwinski MK, Goodman RM. (2004). Comparison of crenarchaeal consortia inhabiting the rhizosphere of diverse terrestrial plants with those in bulk soil in native environments. *Appl Environ Microbiol* 70: 1821-1826.

Smith SE, Read DJ. (1997). *Mycorrhizal symbiosis.* 2nd edn. Academic Press: London, UK. 605 pp.

Smith KS, Ingram-Smith C. (2007). *Methanosaeta*, the forgotten methanogen? *Trends in Microbiology,* 15: 150-155.

Söderström B, Finlay RD, Read DJ. (1988). The structure and function of the vegetative mycelium of ectomycorrhizal plants IV. Qualitative analysis of carbohydrate contents of mycelium interconnecting host plants. *New Phytol* 109: 163-166.

Timonen S, Bomberg M. (2009). Archaea in dry soil environments. *Phytochem Rev* 8: 505-518.

Timonen S, Hurek T. (2006). Characterization of culturable bacterial populations associating with *Pinus sylvestris-Suillus bovinus* mycorrhizospheres. *Can J Microbiol* 52: 769-778.

Timonen S, Tammi H, Sen R. (1997). Characterization of the host genotype and fungal diversity in Scots pine ectomycorrhiza from natural humus microcosms using isozyme and PCR-RFLP analyses. *New Phytol* 135: 313-323.

Timonen S, Jørgensen KS, Haahtela K, Sen R. (1998). Bacterial community structure at defined locations of *Pinus sylvestris Suillus bovinus* and *Pinus sylvestris Paxillus involutus* mycorrhizospheres in dry pine forest humus and nursery peat. *Can J Microbiol* 44: 499-513.

Tourna M, Stieglmeier M, Spang A, Könneke M, Schintlmeister A, Urich T, Engel M, Schloter M, Wagner M, Richter A, Schleper C. (2011). *Nitrososphaera viennensis*, an ammonia oxidizing archaeon from soil. *Proc Nat Acad Sci USA* 108: 8420-8425.

Treusch AH, Leininger S, Kletzin A, Schuster SC, Klenk HP, Schleper C. (2005). Novel genes for nitrite reductase and Amo-related proteins indicate a role of uncultivated mesophilic crenarchaeota in nitrogen cycling. *Environ Microbiol* 7: 1985-1995.

Tugel A, Lewandowski A, Happe-vonArb D. (2000). Soil Biology Primer [online]. Available: soils.usda.gov/sqi/concepts/soil_biology/biology.html [12.03.2008]

van Hees,Patrick A. W., Rosling A, Essen S, Godbold DL, Jones DL, Finlay RD. (2006). Oxalate and ferricrocin exudation by the extramatrical mycelium of an ectomycorrhizal fungus in symbiosis with *Pinus sylvestris*. *New Phytol* 169: 367-378.

Wallander H, Nilsson LO, Hagerberg D, Bååth E. (2001). Estimation of the biomass and seasonal growth of external mycelium of ectomycorrhizal fungi in the field. *New Phytol* 151: 753-760.

Wang XP, Zabowski D. (1998). Nutrient composition of Douglas-fir rhizosphere and bulk soil solutions. *Plant and soil* 200: 13-20.

Woese CR and Fox GE. (1977). Phylogenetic Structure of the Prokaryotic Domain: The Primary Kingdoms. *Proceedings of the National Academy of Sciences* 74: 5088-5090.

Woese CR, Kandler O, Wheelis ML. (1990). Towards a natural system of organisms: proposal for the domains Archaea, Bacteria, and Eucarya. *Proc Natl Acad Sci U S A* 87: 4576-4579.

Xavier LJC, Germida JJ. (2003). Bacteria associated with *Glomus clarum* spores influence mycorrhizal activity. *Soil Biol Biochem* 35: 471-478.

Yang Y, Cui H, Zhou P, Liu S. (2006). *Halobacterium jilantaiense* sp. nov., a halophilic archaeon isolated from a saline lake in Inner Mongolia, China. *Int J Syst Evol Microbiol* 56: 2353-2355.

Yavitt JB, Fahey TJ, Simmons JA. (1995). Methane and carbon-dioxide dynamics in a dorthern hardwood ecosystem. *Soil Sci Soc Am J* 59: 796-804.

Yrjälä K, Katainen R, Jurgens G, Saarela U, Saano A, Romantchuk M, Fritze H. (2004). Wood ash fertilization alters the forest humus Archaea community. *Soil Biol Biochem* 36: 199-201.

Zhuang X, Chen J, Shim H, Bai Z. (2007). New advances in plant growth-promoting rhizobacteria for bioremediation. *Environment International*, 33: 406-413.

In: Spruce ISBN 978-1-61942-494-4
Editors: K. I. Nowak and H. F. Strybel © 2012 Nova Science Publishers, Inc.

Chapter 3

REVERSIBLE VARIATIONS IN SOME WOOD PROPERTIES OF NORWAY SPRUCE (*PICEA ABIES* KARST.), DEPENDING ON THE TREE FELLING DATE

Ernst Zürcher*, Christian Rogenmoser, Azadeh Soleimany Kartalaei, and Diane Rambert
Bern University of Applied Sciences BFH / Architecture,
Wood and Civil engineering Biel, Switzerland

ABSTRACT

Traditional knowledge, in the form of so-called rural rules, indicates that the date of tree felling has an important influence on wood quality. The main factor, after the season of the year, is said to be the position of the moon. The object of the research presented here was to study the variability of some user-related properties of wood, by analyzing measurable parameters. The material stems from four different Swiss sites and is representative of central European conditions. The study involved 576 trees — Norway Spruce (*Picea abies* Karst.) and Sweet Chestnut (*Castanea sativa* Mill.) — felled on 48 dates throughout the fall

---

* Address of the senior author: Ernst Zürcher, Dr. sc. nat., Forest Engineer ETHZ Professor for Wood Science / Research Division Wood Engineering Bern University of Applied Sciences BFHArchitecture, Wood and Civil Engineering AHB, Solothurnstrasse 102, CH-2500 Biel 6, Email: ernst.zuercher@bfh.ch.

and spring of 2003–2004 (always on Mondays or Thursdays). Before the start of the experiment, one sample was taken on the same day from each of the tested trees, to serve as reference. Wood properties analyzed are: water-loss, shrinkage under controlled drying, air dry and oven dry density. The statistical analysis of the complete data series reveals (in addition to a seasonal trend) a generally weak, but highly significant role of the synodic and sidereal moon cycles and, to a lesser extent, the tropical cycle.

The lunar-related differences are more marked for the middle months of the trial. The most obvious variation in Spruce occurs between samples of trees felled immediately before the Full Moon and the samples immediately following Full Moon. Smaller series of Spruce samples were tested on hygroscopicity, compression strength and calorimetry. Here too, the strong value shifts observed around the Full Moon found a clear confirmation. The main variation factor for water uptake is however the type of forest and the site, a naturally grown mountain forest producing a clearly less hygroscopic wood. The results from this study bring some transparency and objectivity into a mainly unexplored field of traditional knowledge, a field subject to controversial discussions. Further research in chronobiology of wood could lead to an ecological technique enhancing specific wood properties.

Keywords: Wood properties, Forestry traditions, Moon phases, Drying, Water sorption, Compression strength, Calorimetry.

RECENT DEVELOPMENTS

The idea of a relationship between plant growth and moon cycles has often been considered by scientists as due to old superstitions. An interesting corpus of well documented experimental work exists and suggests objective phenomena interacting at different levels. An extensive review of the available data and interpretations has recently been published in a book on botany (Zürcher 2008, 2011), based on a series of 88 scientific publications related to this topic.

Since we started working on the field of tree chronobiology linked to lunar rhythms, it has been possible to observe significant relationships for different aspects of tree life and wood properties. Here is a short overview:

- The germination and initial growth of some tropical trees show a decided rhythmic character. Speed of germination, percent of

germination, average height, and maximum height after 4 months are systematically related to the timing of sowing in relation to the moon phase (Zürcher 1992, 2000).
- An interdisciplinary reworking of previously published, long-term tree-physiological research results (variations of tree diameters obtained by extensometry) has enabled researchers to consider an unexpected aspect: the synodic (time required for the moon to complete a full phase, i.e., usually 29.53 days) moon-rhythm at a daily level (gravimetric tide-rhythm) could be established for trees held under constant conditions (darkness) (Zürcher, Cantiani, Sorbetti-Guerri and Michel 1998).
- Data of trees measured in open conditions, reanalyzed recently with more sophisticated tools, brought spectacular confirmation of the role of lunar tides in tree physiology (Barlow, Mikulecky and Strestik 2010). In the meantime, it was possible to detect the same type of fluctuations by measuring with a high-sensitivity device the low-potential electric currents along the trees' stems, depending on the physiological phase of the trees (Holzknecht and Zürcher 2006).
- The drying behavior (water loss/shrinkage) and the final density of wood systematically and coherently vary in function of the tree felling date, if analyzed in relation to the season and to the position of the moon (Zürcher and Mandallaz 2001). The observed fluctuations are yet more complex than mentioned in forestry traditions existing all over the world (Zürcher, Schlaepfer, Conedera and Giudici 2010).

Traditional knowledge and practices concerning agricultural and forestry activities - so-called rural rules - are still widespread in various cultures on different continents. Among them, rules exist about effects of the tree-felling date on the properties of wood (Hauser 1973, Broendegaard 1985, Oldeman 1999, Cole and Balik 2010). The first written evidence of this knowledge dates back to Theophrastus of Eresos (372-287 BC), who in his History of Plants (V, 1, 3) states that there is an appropriate season for cutting the trees and – within the season - if cutting at the beginning of the waning moon, the wood is harder and less likely to rot. This popular knowledge has been passed down to our times and to the local practices of felling trees during different moon positions depending on the specific forms of wood utilization (Zürcher 2000). Despite such a broad and antique tradition in referring to the moon phases for determining the most suitable date for agricultural (e.g. seeding) and forestry (e.g. tree-felling) activities, there are relatively few scientific research works on this topic. For an extensive scientific review about lunar periodicities in Biology see Endres and Schad (1997).

In wood physics, the wood-water relations have been extensively studied (Skaar 1988; Navi and Heger 2005), but not yet including "time" as a possible initial, rhythmic factor of variation in the sample properties; the role of this factor appears at present mainly in the sorption hysteresis and in the viscoelastic behavior. Two publications can nevertheless be cited in relation to the apparently fluctuating plant-water relations at issue, where the time factor "Moon" plays an essential role in living vegetals. Experimental based studies with Bean seeds (*Phaseolus vulgaris*) exist on the lunar cycle dependent water uptake during immersion by a dormant reproductive material (Brown and Chow 1973, Spruyt et al. 1987).

Concerning the role of the lunar phase at the felling date, three almost simultaneous, geographically independent research studies investigated selected wood properties of Norway Spruce with 120 trees in Dresden (Triebel 1998), 60 trees in Freiburg i.Br. (Seeling and Herz 1998, Seeling 2000) and 30 trees in Zürich (Rösch 1999, Bariska and Rösch 2000) respectively. However, none of these three investigations, with 6 felling dates each, could at first significantly confirm the influence of the moon-correlated felling date on the wood properties tested.

The data collected from the 30 trees by Rösch (1999) were reanalyzed, regarding oven-dry wood density as a dependent variable in the same sense as water-loss and shrinkage (Zürcher and Mandallaz 2001, Zürcher 2003). This approach led to encouragingly significant results on the criteria water-loss, shrinkage and relative density (oven-dry density in percent of initial fresh / "green" density).

LARGE SCALE RESEARCH

In order to fill the existing gaps in the choice of the previous experimental felling dates and to obtain a suitable data set for a broader statistical analysis, we set up a large (in terms of number of trees, experimental sites and felling dates) and systematic (always at the same days of the week) tree-felling experiment. The aim was to test the existence of a physical phenomenon of lunar-related variations in some selected wood properties as suggested by the ancient, traditional rules and practices.

Figure 1. Preparation of test samples from Spruce (left: sapwood / right: outer heartwood).

In particular, we wanted to examine the occurrence of rhythmic, time-dependent cyclic variations in the wood-water-relation in correlation with three different types of moon rhythms or cycles. Moreover, different tree species (Norway Spruce and Sweet Chestnut) and different wood types (sapwood and heartwood) within the tree were analyzed separately.

The wood material tested comes regularly and simultaneously from four different Swiss sites representative for Central European conditions. Per main site, 144 Norway Spruce (*Picea abies* Karst.) trees or Sweet Chestnut (*Castanea sativa* Mill.) trees have been felled. These 576 trees were felled at 48 dates (06.10.2003 - 18. 03.2004, always on Mondays and Thursdays). Before the start of the experiment, one reference sample had been taken in the same day from each of the later felled trees (prismatic sample at breast height). Spruce samples were taken both from sap- and outer heartwood (Figure 1), Chestnut samples from heartwood only – the sapwood zone being too narrow in this species. This report presents the obtained variations in values of water loss under controlled drying, relative density (ratio oven-dry density / fresh density) and hygroscopicity of previously dried samples for the main studied species - Norway Spruce.

The data are reported here in a descriptive and graphical way, the statistical analysis being presented in Zürcher, Schlaepfer, Conedera and

Giudici (2010). To illustrate the discovered lunar-related phenomena, we will concentrate on one representative site (Château d'Oex) and one type of wood (sapwood), adding the results of mechanical tests and analysis of energy content (calorimetry) on specific smaller series.

In all cases, strong time-dependent and inter-dependent variations were observed. According to variance analyses based on the totality of data (4032 values per criterium), a highly significant part of the changes in the values for water loss, shrinkage and relative density (oven-dry density green density) occurs in tune with the moon phases and its astronomical positions (synodic, tropic and sidereal rhythms).

The implication of the time-independent reference-density of each tree allows us to confirm the significance of the tested lunar models and to estimate their respective explicative power.

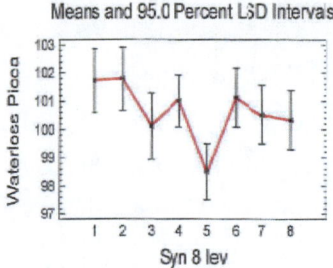 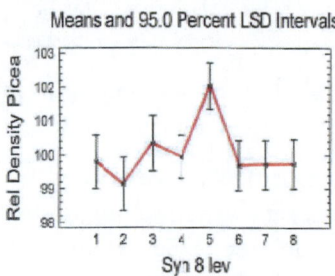

Figure 2. Lunar synodic variations of Water Loss (A) and Relative Density (B) of Norway Spruce (sapwood and heartwood), when dried from "green" to oven-dry state. Mean values with 95% least significant differences. The vertical bold lines indicate the times of full moon, the dotted lines the times of new moon.

If the synodic lunar month (4 quarters) is subdivided in 8 periods of about 3.5 days each ("syn1" to "syn 8"; "syn 1" beginning with the New Moon NM, "syn 5" beginning with the Full Moon FM), one of the strongest "jumps" is occurring around FM, from "syn4" to "syn 5". The clearly opposite variations in Waterloss (Figure 2 A) and Relative Density (Figure 2B) illustrate the importance of the wood-water interface, which obviously undergoes lunar-correlated reversible changes. The variance analysis relatively to the sidereal Moon cycle (position of the Moon in the zodiacal constellations) showed very high levels of significance as well, compared to the synodic cycle.

Therefore, these results confirm the existence of the "Moon" factor as mentioned in traditional knowledge, but in a much more complex form

(Theophrastus' rule seeming to be nearest to the synodic phenomena observed here on Spruce).

FOCUSING ON ONE REPRESENTATIVE SITE

These systematic variations in the course of the synodic lunar month can be illustrated on the variation curve of the criteria "Relative Density" (Figure 3) and Water Loss (Figure 4) among the 48 successive felling dates, for a representative series: Sapwood samples from the site Château d'Oex (even-aged planted mountain Spruce stand). The following table indicates what fellings occurred directly before and directly after Full Moons, with the 6 corresponding dates (Table 1).

Table 1. Tree fellings around Full Moons, and their exact dates

Felling dates (before FM / after FM)	Days of Full Moon (FM)
2 / 3	October 10, 2003
10/11	November 9, 2003
19/20	December 8, 2003
27/28	January 7, 2004
36/37	February 6, 2004
44/45	March 7, 2004

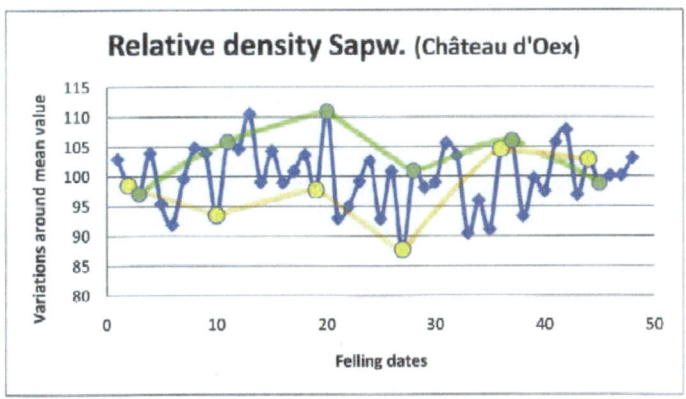

Figure 3. Variation curve of Relative Density of sapwood samples of Spruce from Château d'Oex. Legend of fellings around Full Moon: O = fellings in the 3.5 days before FM / O = fellings in the 3.5 days after FM. Each point represents 12 values (here without std.dev.).

Table 2. Values of Relative Density in the 3.5 days before Full Moon (bFM), compared to the values in the 3.5 days after (aFM) and their difference, for the winter period November to February

Relative density (Sapwood)							
Variation from value before FM (syn 4) to value after FM (syn 5)							
	Oct.	Nov.	Dec.	Jan.	Feb.	Mar.	Mean [%]
bFM (syn4)	-	93.5	97.7	87.7	104.6	-	
aFM (syn5)	-	105.8	110.9	100.9	106	-	Increase
Variation [%]	-	12.3	13.2	13.2	1.4	-	+ 10.0

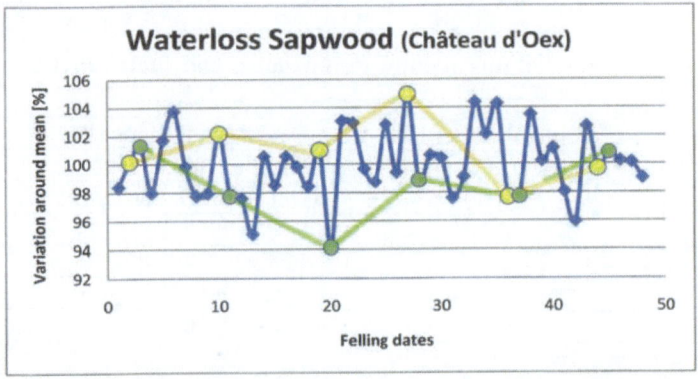

Figure 4. Variation curve of Water Loss of sapwood samples of Spruce from Château d'Oex. Legend of fellings around Full Moon: O = fellings in the 3.5 days before FM / O = fellings in the 3.5 days after FM. Each point represents 12 values (here without std.dev.).

As can be observed, during the four winter months, from November (felling 10/11) to February (felling 36/37), the values before FM are always lower than the ones after. Before and after this core period in the first and the last weeks of the trial (October and March), the variations do not seem to follow the same lunar-related rule. The corresponding values for the middle 4 months around the global site mean (100%) are given in the Table 2, with their differences and an estimation of the mean increase in Relative Density (10%).

The cause for these variations in Relative Density lies in similarly fluctuating values in the loss of free and bound water during the controlled drying process, but in an opposite sense (Figure 4 / compare also Figure 2A with 2B). Here too, marked differences around Full Moons occur in middle winter months only. Table 3 indicates the respective variations around the FM during this period, the general mean being a reduction in Water Loss of 4.5%.

Table 3. Values of Water Loss in the 3.5 days before Full Moon (bFM), compared to the values in the 3.5 days after (aFM) and their difference, for the winter period November to February

Waterloss (Sapwood)							
Variation from value before FM (syn 4) to value after FM (syn 5)							
	Oct.	Nov.	Dec.	Jan.	Feb.	Mar.	Mean [%]
bFM (syn4)	-	102.2	101	104.9	97.7	-	
aFM (syn5)	-	97.7	94.2	98.3	97.7	-	Reduction
Variation [%]	-	-4.5	-6.8	-6.6	0	-	- 4.5

TESTS ON WATER SORPTION (AFTER DRYING)

One of the most important physical properties of wood, decisive for weathering behavior and for decay resistance is its hygroscopicity. This feature is usually expressed as Equilibrium Moisture Content EMC under a given atmosphere, or as Fiber Saturation Point FSP, below which the loss of bound water in the drying process provokes shrinkage. The test method chosen for estimating the expected variations of hygroscopicity was to expose the samples to direct contact with water. One series was made of air-dried and conditioned sticks with dimensions of 16*7*65 mm (see Figure 1). These were all fixed together (12 * 48 = 576 samples per site) at one end to a plate, with the other end being immersed 5 mm deep for 9 minutes in a basin of water colored with ink. The capillar ascent of water was estimated by calculation of the relative weight increase. Figure 5 shows the variations of water sorption by capillarity in samples collected during this period of 24 weeks (48 tree fellings). Both the limitation of the "lunar effect" on the 4 winter months and the systematic drop directly after the FM is very similar to what was observed on initial Water Loss. A noticeable fact is that for capillarity, the mean amplitide of this decrease (25.9%) is much more pronounced (see Table 4).

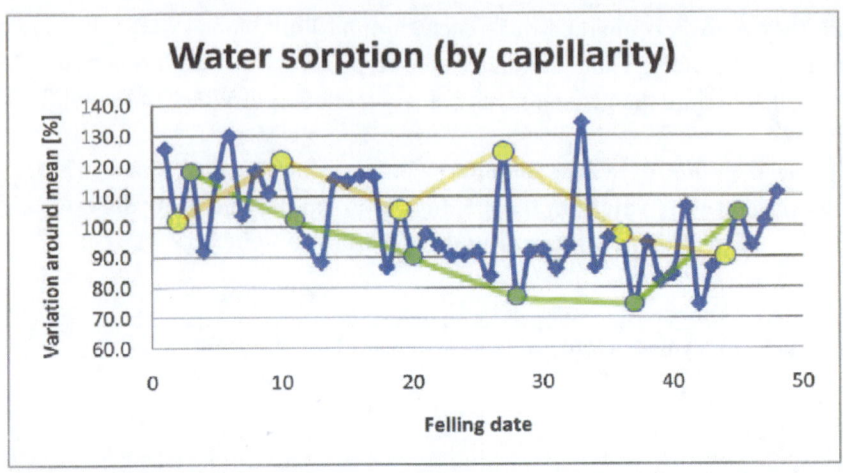

Figure 5. Variation curve of capillar Water Sorption of sapwood samples of Spruce from Château d'Oex. Legend of fellings around Full Moon: O = fellings in the 3.5 days before FM / O = fellings in the 3.5 days after FM. Each point represents 12 values (here without std.dev.).

Table 4. Values of capillar Water Sorption in the 3.5 days before Full Moon (bFM), compared to the values in the 3.5 days after (aFM) and their difference, for the winter period November to February

Water sorption (by capillarity)							
Variation from value before FM (syn 4) to value after FM (syn 5)							
	Oct.	Nov.	Dec.	Jan.	Feb.	Mar.	Mean [%]
bFM (syn4)	-	121.7	105.3	124.6	96.3	-	Reduction
aFM (syn5)	-	102.4	90.3	77	74.5	-	
Variation [%]	-	-19.3	-15	-47.6	-21.8	-	-25.9

The second test method for hygroscopicity was applied by immersion of the samples previously used for density determination. In a similar way, 576 cubic samples for each site (4 per tree, 12 per felling date) were dipped into water (20°C) for 7 days (168 hours).

The sorption is given by the mass percentage increase. The variation curve (Figure 6) is represented relative to the general mean of the whole series. Here too, the systematic lunar variations of sorption occur in obvious coherence with the Water Loss, and inversely to the Relative Density. Table 5 indicates that the mean reduction in sorption from the days before to the days after the Full Moon is 12.6%, half the value obtained for capillar water uptake.

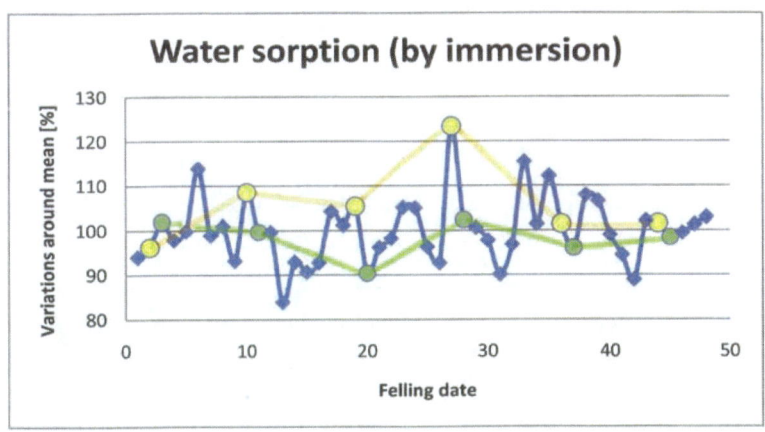

Figure 6. Variation curve of Water Sorption by immersion of sapwood samples of Spruce from Château d'Oex. Legend of fellings around Full Moon: O = fellings in the 3.5 days before FM / O = fellings in the 3.5 days after FM. Each point represents 12 values (here without std.dev.).

Table 5. Values of Water Sorption by immersion in the 3.5 days before Full Moon (bFM), compared to the values in the 3.5 days after (a FM) and their difference, for the winter period November to February

Water sorption (by immersion)							
Variation from value before FM (syn 4) to value after FM (syn 5)							
	Oct.	Nov.	Dec.	Jan.	Feb.	Mar.	Mean [%]
bFM (syn4)	-	108.5	105.4	123.5	101.5	-	
aFM (syn5)	-	99.7	90.4	102.4	96.1	-	Reduction
Variation [%]	-	-8.8	-15	-21.1	-5.4	-	-12.6

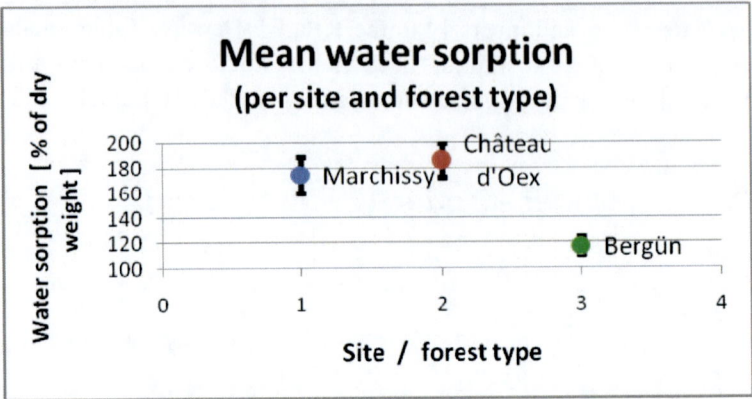

Figure 7. Mean water sorption after 7 days immersion (in % of dry weight, with standard deviations) of samples from Spruce plantations (Marchissy, sapwood / Château d'Oex, sapwood), compared to samples from a naturally grown mountain forest (Bergün, heartwood). Each mean value represents 576 samples.

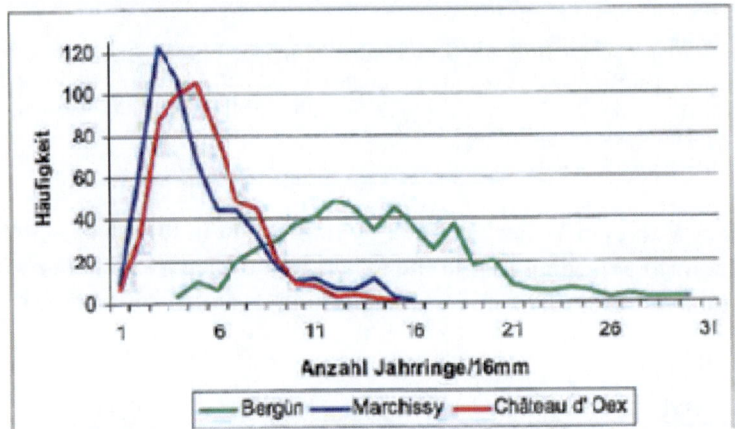

Figure 8. Distribution of growth ring numbers on 16 mm-samples from 2 plantation sites (Marchissy and Château d'Oex) and one natural mountain forest (Bergün). X-axis: Number of growth rings per 16 mm-samples / Y-axis: Frequency.

It must be stressed that lunar or seasonal variability in water sorption plays a secondary role compared to the much more pronounced effect of forest type, site and sample provenance in the stem. Wood from naturally grown alpine forests (Bergün, 1550 m) was compared to wood from planted Spruce trees in an alpine site as well (Château d'Oex, 1200 m) as from a planted stand in low altitude (Marchissy, 650 m), in terms of absolute values of water sorption during 7 days of immersion (all samples having been previously dried at the same level).

For this comparison, heartwood samples have been used from the Bergün material (slow-grown mountain timber showing a narrow peripheral sapwood zone) and sapwood samples were used from the fast-grown plantation wood of Château d'Oex and Marchissy (where this outer zone is much wider and usually present in parts of sawn products).

The difference in hygroscopicity is considerable - the Bergün samples absorbing 43% less than the Marchissy samples and 46% less than the Château d'Oex samples (Figure 7).

Additional tests with naturally grown alpine wood have shown that this difference is not due to the fact that the Bergün samples came from the heartwood zone: in such a situation, the permeability of sapwood is only slightly higher than the one of heartwood.

Possible differences in density do not explain this exceptional hygroscopic behavior, all three sample series having a similar low weight: the oven-dry density of the Bergün samples was 0.401 g/cm3 (Stdev 0.018), the one of the Marchissy samples 0.398 g/cm3 (Stdev 0.032) and Château d'Oex 0.379 g/cm3 (Stdev 0.021). As an expression of the respective growth dynamics, the Marchissy and Château d'Oex samples have a similar growth ring configuration (generally fewer rings per sample), compared to Bergün, with significantly more rings per sample (Figure 8). Despite the fact that a certain number of Bergün samples were wide-ringed and some Marchissy / Château d'Oex samples narrow-ringed, there was no such superposition in their hygroscopicity (Figure 7). This underlines clearly the importance of the forest type and sylviculture at the origin of a strongly different sorptive behavior (in immersion as well as by capillarity) - wood from natural mountain forests showing an unexpected advantage (presented here for the first time).

COMPLEMENTARY TESTS ON COMPRESSION STRENGTH

Complementing the experiments on hygroscopic behavior of Spruce wood, mechanical tests have been performed in order to determine a representative feature: the compression strength parallel to the grain. Samples of transverse section 20 mm x 20 mm have been tested according to the current DIN/EN standards with a universal test machine "Zwick/Roell" (2050).

Out of the complete test series dealing with 548 samples, the values corresponding to the tree felling dates around the Full Moons (6 x 2 = 12 dates), based on 132 samples, are presented here. This allows direct comparisons with the lunar-periodic reversible variations in density and in hygroscopicity. A similar fluctuation can be observed here as for the Relative Density (Figure 9, to be compared with Figure 3). Additionally, the same mean amplitude of increase is calculated for the 4 winter months (Table 6), the global difference being significant too (P-value = 0.004).

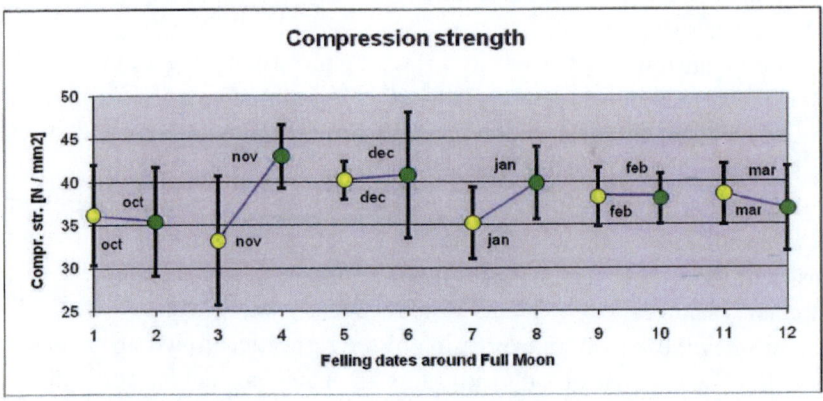

Figure 9. Variations of compression strength parallel to the grain of sapwood /outer heartwood samples of Spruce from Château d'Oex. Legend of fellings around Full Moon: 1, 3, 5, 7, 9, 11, marked by O , for fellings in the 3.5 days before FM / 2, 4, 6, 8, 10, 12, marked by O , for fellings in the 3.5 days after FM. Each point represents in average 11 values (here with corresponding std.dev.).

Table 6. Values of Compression strength parallel to the grain in the 3.5 days before the Full Moon (bFM), compared to the values in the 3.5 days after (aFM) and their difference, for the winter period November to February

Compression strength (in N / mm2, parallel to the grain)							
Variation from value before FM (syn 4) to value after FM (syn 5)							
	Oct.	Nov.	Dec.	Jan.	Feb.	Mar.	Mean [%]
bFM (syn4)	-	33.2	40.2	35.2	38.2	-	
aFM (syn5)	-	43.0	40.8	39.8	38.0	-	Increase
Variation [%]	-	+ 29.5	+ 1.5	+ 13.1	- 0.5	-	+ 10.9

COMPLEMENTARY TESTS ON CALORIMETRY

The calorific value is another important physical feature of wood, decisive for its energetic use. This value is usually expressed per unit of weight, with an average amount of ca. 19 MJ/kg for absolutely dry wood (Fengel and Wegener 2003). In a more practical way, the energy content of wood can also be determined per unit of volume (the form in which commercialization occurs), taking into account differences in density inside one species, but also between species.

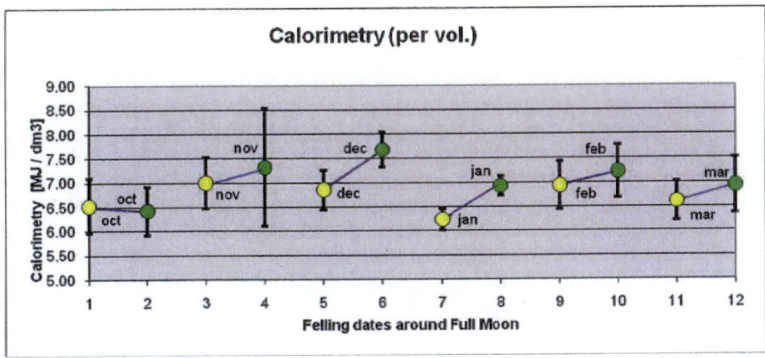

Figure 10. Variations of calorific values per volume of sapwood /outer heartwood samples of Spruce from Château d'Oex. Legend of fellings around Full Moon: 1, 3, 5, 7, 9, 11, marked by O , for fellings in the 3.5 days before FM / 2, 4, 6, 8, 10, 12, marked by O , for fellings in the 3.5 days after FM. Each point represents in average 9 values (here with corresponding std. dev.).

Table 7. Values of calorimetry in the 3.5 days before Full Moon (bFM), compared to the values in the 3.5 days after (aFM) and their difference, for the winter period November to February

Calorimetry (in MJ / dm3)							
Variation from value before FM (syn 4) to value after FM (syn 5)							
	Oct.	Nov.	Dec.	Jan.	Feb.	Mar.	Mean [%]
bFM (syn4)	-	7.00	6.85	6.24	6.94	-	
aFM (syn5)	-	7.32	7.68	6.93	7.23	-	Increase
Variation [%]	-	+ 4.6	+ 12.1	+ 11.1	+ 4.2	-	+ 8.0

A number of 102 Spruce sapwood samples of cubic shape ca. 16 mm x 16 mm x 16 mm in size have been tested according to the current ISO 1716 standards with a calorimetry device from Fire Testing Technology U.K., using a "calorimetric bomb" from Parr Instrument Company, Illinois, USA. For each of the felling dates around the Full Moons (6 x 2 = 12 dates), 3 samples from 3 trees (9 samples in total) were tested (in two cases, only 6 samples were available), which made a total of 102 calorimetric tests (Rambert 2011). While the mean calorific value per unit of weight was 19.92 MJ/kg, with a standard deviation of 0.25 - representing a variation coefficient of only 1% - the mean calorific value per unit of volume was 6.90 MJ/dm3, with a standard deviation of 0.67 - representing a considerable variation coefficient of 10%. A similar fluctuation can be observed here as for the Relative Density (Figure 10, to be compared with Figure 3). Additionally, an analogous mean amplitude of increase is calculated for the 4 winter months (Table 7), the global difference for the months November - February being highly significant here as well (P-value = 0.002).

CONCLUSION

From these results and observations, following 4 points can be stressed:

- Moon-related reversible variations are not limited to Water Loss, Relative Density (and Shrinkage) during the drying process: they are an even more marked phenomenon at the level of water sorption (capillary and by immersion) by previously dried wood samples
- The variations in hygroscopicity (water sorption) are however much weaker than the differences due to the site and to the type of sylviculture, samples from a naturally grown mountain forest showing a much lower water uptake than samples from planted Spruce stands
- The mechanical and calorimetric tests show moon-related reversible variations as well, consistent with the variations in density. It must be stressed that in addition to the calorific value, combustibility (ability to burn easily) plays an important role in energetic use of wood. Specific laboratory tests on samples of Château d'Oex showed similar variations around the Full Moon as for Waterloss, but in a much lower amplitude. This result would be consistent with the traditional rule

that good firewood is obtained during the waxing Moon (before Full Moon)
- These systematic and coherent lunar rhythmicities occur clearly only during the 4 months November to February, which indicates that both season and lunar cycles are involved
- These encouraging results need to be confirmed by durability tests before drawing a definitive conclusion in the sense of general outdoors applicability for Spruce. The plausible difference in terms of decay resistance of Spruce wood can nevertheless be supposed as follows: lower durability from fellings shortly before Full Moon - higher durability from fellings immediately after Full Moon between November and February.

Table 8 gives a general overview of the obtained results:

Table 8. Changes in the presented criteria, for samples of Spruce wood from the 3.5 days before Full Moon, compared to the values of samples from the 3.5 days after FM

Comparative changes around Full Moon (in the months November, December, January, February)		
Features	Samples from 3 ½ days before Full Moon (syn.4)	Samples from 3 ½ days after Full Moon (syn.5)
Relative Density	lower	Higher
Water Loss	higher	Lower
Water Sorption (by capillarity)	higher	Lower
Water Sorption (by immersion)	higher	Lower
Compression strength	lower	Higher
Calorific value (per unit of volume)	lower	Higher

ACKNOWLEDGMENTS

This research was supported by funds provided by Wolfermann-Nägeli-Foundation Zürich, Chambre des Bois de l'Ouest Vaudois, Sezione Forestale Cantonale (Ticino), Federlegno Ticino, Graubünden Holz (Amt für Wald

Mittelbünden), Kantonsforstamt Schwyz, Schwyzer Arbeitsgemeinschaft HOLZ, Kloster Einsiedeln SZ and Thoma Holz GmbH (Austria). The realization occurred with the collaboration of Dr. Thomas Volkmer, Werner Gerber, Andrea Florinett, Christian Barandun, Eric Treboux, Denis Pidoux, Serge Lüthi, Christophe Rémy, Daniel Meyer, Marco Delucchi, Theo Weber, and Dr. Joan Davis.

REFERENCES

Bariska, M., Rösch, P. 2000: Felling Date and Shrinkage Behaviour of Norway Spruce. (In German, English summary) Schweiz. Z. Forstwes. 151 (2000) 11: 439-443.

Barlow, P.W., Mikulecky M., Strestik J. 2010: Tree-stem diameter fluctuates with the lunar tides and perhaps with geomagnetic activity. *Protoplasma* (online): April 15, 2010.

Broendegaard, V.J. 1985: Ethnobotany: Plants in traditions, history and popular medicine - Tree felling and moon phases: superstition or folk-visdom? (In German) In: *Contributions to Ethnomedicine, Ethnobotany and Ethnozoology.* Verl. Mensch und Leben, Berlin, Bd. 6: 82-92.

Brown, F.A., Chow, C.S. 1973: Lunar-correlated variations in water uptake by bean seeds. *Biol.Bull.* 145: 265-278.

Cole, I.B., Balik, M.J., 2010: Lunar Influence: Understanding Chemical Variation and Seasonal Impacts on Botanicals. American Botanical Council. Herbal Gram. 2010;85:50-56.

Endres, K.-P., Schad, W. 1997: Biology of the Moon. Moon periodicities and Life Rhythms. (in German) S.Hirzel Verlag, Stuttgart / Leipzig, 308 pp.

Fengel, D., Wegener, G. 2003: *WOOD – Chemistry*, Ultrastructure, Reactions. Verlag Kessel, Remagen.

Hauser, A. 1973: Rural rules. *A Swiss collection with comments.* (In German) Artemis Verlag, Zürich, München, 710 pp.

Holzknecht K., Zürcher E. 2006: Tree stems and tides – A new approach and elements of reflexion. *Schweizerische Zeitschrift für Forstwesen.* 2006;157(6):185-190.

Navi, P., Heger, F. 2005: Thermo-hydromechanical behavior of wood. (In French) Presses polytechniques et universitaires romandes, *Lausanne,* 298 pp.

Oldeman, R.A.A. 1999: Personal communication. Prof.dr.ir., *LUW - Hutan Lestari Team,* Wageningen.

Rambert, D. 2011: Propriétés hygroscopiques, mécaniques et calorifiques du bois d'Epicéa: étude de facteurs de variation. Bachelor Thesis, Bern University of Applied Sciences / Architecture, *Wood and Civil engineering,* Biel, Switzerland.

Rösch, P. 1999: Research on the influence of the moon phase-related felling date on the drying process and shrinkage of Norway Spruce-wood (Picea abies Karst.). (In German, English summary) Diploma thesis, Swiss Federal Institute of Technology, *Wood Sciences,* Zürich, 42 pp.

Seeling, U. 2000: *Selected Wood Properties of Norway Spruce (Picea abies L. Karst) and its Dependance on the Date of Felling.* (In German, English summary) Schweiz. Z. Forstwes. 151 (2000) 11: 451-458.

Seeling, U., Herz, A. 1998: Influence of felling date on shrinkage and water content of Norway Spruce-wood (Picea abies Karst.). A literature survey and pilot research. (In German) Albert-Ludwigs-University, *Forest Sciences,* Freiburg i.Br., 98 pp.

Skaar, C. 1988: *Wood-Water Relations.* Springer-Verlag, Berlin, Heidelberg, New York, London, Paris, Tokyo, 283 pp.

Spruyt, E., Verbelen, J.-P., De Greef, J. A. 1987: Expression of Circaseptan and Circannual Rhythmicity in the Imbibition of Dry Stored Bean Seeds. *Plant Physiol.* 84: 707-710.

Triebel, J. 1998: Moon phase-dependent tree-felling - A literature survey and research on some properties of Norway Spruce (Picea abies Karst.). (In German) Technical University of Dresden, *Forest Sciences,* Tharandt, 108 pp.

Zürcher, E. 1992: Rhythmizitäten in der Keimung und im Initialwachstum einer tropischen Baumart. Rythmicités dans la germination et la croissance initiale d'une essence forestiere tropicale. *Schweizerische Zeitschrift für Forstwesen.* 1992; 143(12):951-966.

Zürcher, E. 2000: *Moon-Related Traditions in Forestry and Corresponding Phenomena in Tree Biology.* (In German, English summary) Schweiz. Z. Forstwes. 151 (2000) 11: 417-424.

Zürcher, E. 2008: Les Plantes et la Lune – Traditions et Phénomènes. In: *Aux Origines des Plantes – Des plantes anciennes à la botanique du XXIè siècle.* Hallé F., ed. Paris: Fayard; 2008:389–411.

Zürcher, E. 2011: *Plants and the Moon - Traditions and Phenomena.* American Botanical Council, Herbal E Gram Vol. 8 Number 4, April 2011 (14 p.).

Zürcher, E,, Cantiani, M-G., Sorbetti-Guerri, F., Michel, D. 1998: Tree stem diameters fluctuate with tide. *Nature.* 1998;392:665-666.

Zürcher, E., Mandallaz, D. 2001: Lunar Synodic Rhythm and Wood Properties: Traditions and Reality. Experimental Results on Norway Spruce (Picea abies Karst.). Proc. 4th Int. Symp. *Tree Biology and Development.* Isabelle Quentin Publ., Montreal, 244-250.

Zürcher, E., Schlaepfer, R., Conedera, M., Giudici, F.. 2010: Looking for differences in wood properties as a function of the felling date: lunar phase-correlated variations in the drying behavior of Norway Spruce (*Picea abies* Karst.) and Sweet Chestnut (*Castanea sativa* Mill.). *Trees.* 2010;24:31-41.

In: Spruce
Editors: K. I. Nowak and H. F. Strybel © 2012 Nova Science Publishers, Inc.
ISBN 978-1-61942-494-4

Chapter 4

EXPLORATION OF FOREST VEGETATION WITH DYNAMIC MODELING, GIS AND REMOTE SENSING

Lubos Matejicek
Institute for Environmental Studies
Charles University in Prague
Faculty of Natural Science
Benatska 2, 128 01, Prague
Czech Republic

ABSTRACT

Landscape spruce planning operates on the interface between management actions and ecosystems. It requires understanding the existing ecological state of the spruce forest and projecting vegetation over time and space to accomplish management objectives. New processing tools are needed to provide more complex spatio-temporal analysis. Nowadays, Geographic Information Systems (GISs) extended by remote sensing techniques, and terrain measurements with Global Positioning Systems (GPS) offer advanced methods for landscape spruce planning and management. Using satellite images and aerial photographs is presented as a revolutionary approach that is expanded by a next generation of sensors with the improved spectral and spatial resolution. Remote sensing has been a valuable source of information for a few decades in mapping and monitoring forest activities. Exploratory spatial data analysis and dynamic modeling enable to evaluate spatial relationships in soil, water, and wildlife resources. In order to demonstrate advanced processing tools focused on spruce ecology, management and conservation, a case study dealing with the spatio-temporal modeling of natural regeneration is carried out in the GIS that can provide a high-quality spatial database. It describes the state of the experimental areas of interest in the spruce forest, and includes a spatio-temporal model to create alternatives for planning and management. The numerical model is based on a large set of ordinary differential equations that can solve dynamic processes and spatial relationships in selected microsites. The simulation results can show the short-time succession for a regeneration decade and approximate long-term development. The use of GIS offers visualization of model outputs that offer to present the decision-making processes in a more illustrative way.

Keywords: Forest management; dynamic modeling; GIS; remote sensing; GPS.

1. INTRODUCTION

The roots of sustainable forest management have been observed for a few centuries. For a long period of time, people have been interested in extracting immediate value from forests, while preserving their characteristics for future generations. Nowadays, human activities in forests are increasingly monitored using plans that are focused on the sustainability and the preservation of biodiversity [1]. The plans are designed to maintain the long-term health of forest ecosystems, while providing ecological, economic, social, and cultural opportunities for the benefit of present and future generations [2]. In current times, as populations and industry continue to expand, world forests increasingly appear finite and vulnerable. Thus, many efforts have been provided to the issues of sustainable forest management in order to have far-reaching implications for the way in which society will view and use the forest resources of the planet. Nowadays, increased emphasis on scientific management is indicated and the need to explore better ecosystem functioning in a global and local spatial scales [3] likewise, in long and short periods of time [4]. The changes from a traditional forest management approach with an emphasis on timber values to sustainable forest management are profound. It is subject to continuing wide-ranging discussions. The historical and current mismanagement can reach the critical point, after which the damage is overwhelming and final. Destructive land use practices and widespread pollution have caused damage to the Earth's forest ecosystems. In spite of sustainable forest management applied in many areas, there are still many questions about the end results of changing values in society and the social view of forests. It is expected that increasing amounts of scientific information and the implementation of new research technology must be acquired to support the ongoing goals and objectives in managing sustainable forests.

In order to explore forest ecosystems in a more complex way, a suite of new information technologies, including geographic information systems (GIS) [5] are needed to manage terrain observations and data from remote sensing. New information can be obtained by observations on long-term sampling plots, analysis of historical plans, and spatio-temporal models [6]. GISs are able to access data from various purpose-designed national and regional resources such as forest health networks, decision support system networks, and ecological classification systems. After data processing, GIS methods offer more complex analysis from new and existing forest inventories.

2. GIS for Forest Management and Protection

GISs and related information technologies are often considered the domain of geographers, a myth that tends to be perpetuated by geographers and ecologists alike. Putting aside disciplinary preconceptions, the recent GIS can address many scientific needs in forest management and protection. It is best suited to analyze questions of a spatial nature such as in which the location of a biological entity relative to other entities or the environment influences its functioning. The tool-base definition of a GIS is a powerful set of tools for collecting, storing, retrieving at will, transforming and displaying spatial data from the real world for a particular set of purposes [7]. The spatial data represents phenomena from the real world in terms of their positions, attributes and spatial inter-relations. The positions are defined with respect to a known coordinate system that is managed by GIS. The attributes are unrelated to positions and can contain information focused on tree conditions, boundaries of monitored areas, or classification of forest sites. The spatial interactions known as topology describe how objects or processes are linked together. The spatial data are usually managed in the framework of the spatial database that support data models focused on efficient storage of positions, attributes and topology.

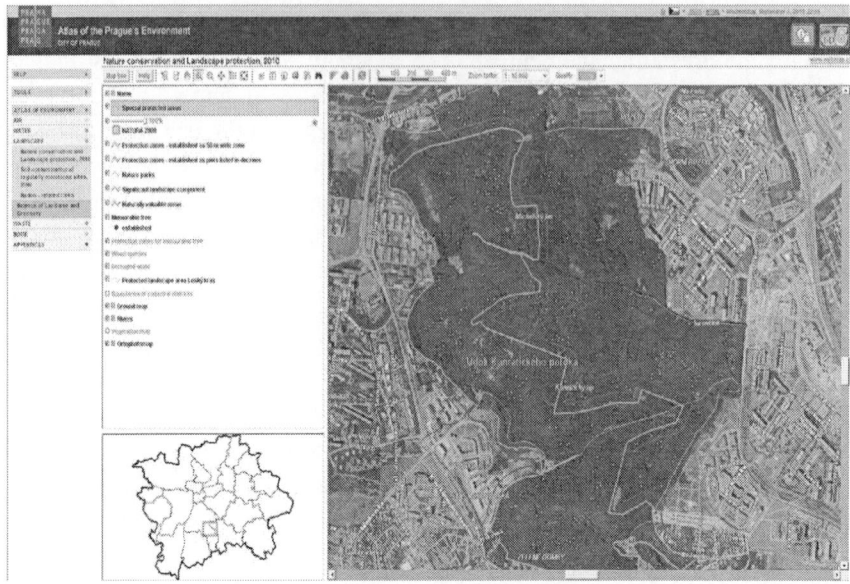

Figure 1.GIS project: Atlas of the Prague's Environment, City of Prague 2011 (http://www.premis.cz/atlaszp/En_default.htm, September 2011).

The spatial data in GIS are mostly represented by points, lines, polygons, and images that are organized into thematic layers. Thematic layers are used for modeling forest sites and processes. The maps combine many thematic layers over a common geographic area. These thematic layers contain collections of similar entities and model their attributes, relationships, and behavior [8]. The similar elements and principles are employed by ArcGIS that uses the geodatabase as the native data structure for the storage and analysis of spatial data. In a similar way like maps contain a collection of many thematic layers, the geodatabase consists of a collection of thematic datasets [9, 10]. It enables to run a wide range of functions focused on exploration, analysis and visualization of spatial data dealing with forest management and conservation [11, 12]. Figure 1 illustrates a part of the spruce forest surrounded by urban areas in the City of Prague. The map layers contain the image of the area of interest, boundaries of the protected zones, wood species extended by memorable trees and other thematic entities. The list of the map layers indicates displayed sets of entities in the GIS environment.

3. REMOTE SENSING FOR FOREST MANAGEMENT AND PROTECTION

GISs are linked with other types of software in order to obtain remote observations, terrain measurements, and database support [13]. Nowadays, processing of satellite images and aerial photographs represents a revolutionary approach that is expanded by a next generation of sensors with the improved spectral and spatial resolution. Remote sensing has been a valuable source of information for a few decades in mapping and monitoring forest activities [14, 15]. Exploring the spruce ecology, management and conservation needs increased amounts and quality of information. It is concerned with the spatial distribution of forest resources within their management area and in the surrounding ecosystems supplemented with the timely acquisition of information on conditions and changes to these resources. It helps to explain small and large impacts associated with changing patterns and processes at different scales in space and time, Figure 2-5.

Figure 2. An image derived from aerial photographs of a part of the spruce forest surrounded by urban areas of the City of Prague.

Figure 3. An image derived from aerial photographs of a part of the spruce forest that provides a detailed view on individual tree species.

Figure 4. An multi-spectral satellite image captured by the ETM+ sensor in May 2000 focused on spruce forests surrounded with agricultural and urban areas, an example is created using the near infrared band (0.76-0.90 μm), the red band (0.63-0.69 μm), and the green band (0.52-0.60 μm), which enables to detect forest boundaries and forest vegetation conditions, the original resolution is 30 meters.

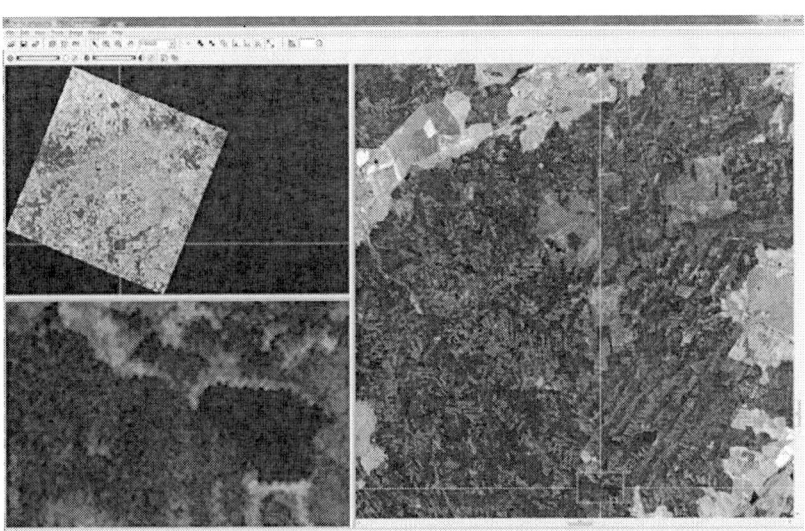

Figure 5. An panchromatic satellite image captured by the ETM+ sensor in May 2000 focused on spruce forests surrounded with agricultural and urban areas, the original resolution is 15 meters (0.52-0.90 μm).

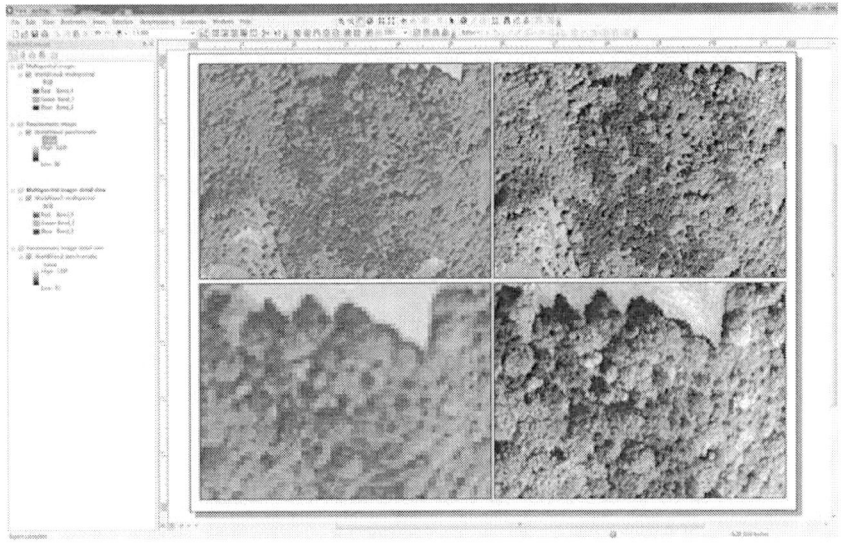

Figure 6. Processing of very high resolution images from WorldView-2 in the ArcGIS environment, the sensors provide 0.46 m panchromatic resolution at nadir (0.450-0.800 μm) and 1.85 m multi-spectral resolution, an example of multi-spectral image is

created using the near infrared band (near-IR1 0.770-0.895 µm), the red band (0.630-0.690 µm), and the green band (0.510-0.580 µm), the panchromatic original resolution is resampled to 0.50m and the multi-spectral original resolution is resampled to 2.0 m to comply with U.S. Regulation.

Remote sensing methods detect electromagnetic energy, which includes such familiar forms as visible light and other parts of the electromagnetic spectrum, mainly near-infrared and far-infrared or thermal-infrared. Some systems also use the microwave portion of the electromagnetic spectrum that includes wavelengths ranging from 1 mm to 2 m penetrating through clouds and atmosphere haze. There are two general types of systems in remote sensing belonging to passive and active devices. Passive remote sensing devices use naturally available energy sources to detect surface features [16, 17, 18, 19].

Photographic cameras in aerial photography are examples of passive sensors. They have been used extensively in forest management since the 1940s. Conventional black and white, color, and color infrared images are acquired over extensive forest regions at a range of scales in support of forest management operations and planning. Progress in photographic science has provided continual improvements in aerial camera technology in order to optimize film speed, contrast, and resolution. Examples in Figure 2-3 illustrate images captured originally with the resolution 0.1 meter. In color and color-infrared films, the different portions of the electromagnetic spectrum expose different layers of yellow, magenta, and cyan dye, which is combined to make red, green, and blue. It can help to detect healthy vegetation due to its higher reflection of the near-infrared electromagnetic spectrum [20, 21, 22].

Satellite imaginaries are created by sensors located on satellite systems launched by research centers such as NASA in the U.S., ESA in Europe and other national agencies. Initially, the major disadvantage of the satellite imaginaries was its relatively poor spatial resolution and higher demands on atmospheric corrections. The first non-military satellite image acquisition program suitable for forest monitoring represents the Landsat series of satellites, started in 1972. They carried Multi-Spectral Scanners (MSS) and subsequently Thematic Mapper (TM) scanners in addition to MSS scanners. The recently operating satellites from Landsat series are Landsat 5 launched in 1985 and Landsat 7 with the enhanced Thematic Mapper (ETM+) launched in 1999. The ETM+ has improved spatial resolution than its Landsat predecessors. In 2003, its scan-line corrector (SLC) failed, resulting in about 22% of the pixels per scene not being scanned [23]. Thus, new compatible sensors are expected to be launched in order to continue with the analysis of temporal changes of forest ecosystems [24, 25, 26]. Besides the detection of visible light in the blue-green-red (RGB) spectral bands, ETM+ has one near-infrared (NIR) band, two mid-infrared bands, and one thermal-infrared (TIR) band [27]. The NIR band is useful for determining vegetation types and biomass content. Examples in Figure 4-5 illustrate images captured originally with the resolution 30 meters for the multi-spectral image and 15 meters for the panchromatic image. In addition to the Landsat series, many satellite systems have been launched for a last few decades such as SPOT [28], TERRA with ASTER and MODIS sensors [29]. The next generation of satellite systems such as Ikonos [30], QuickBird [31] and WorldView-2 [32] includes very high resolution sensors. As an example, the processing of a very high resolution image in the ArcGIS environment is illustrated in Figure 6.

Another category is represented by hyperspectral sensors that can sample the electromagnetic spectrum in many narrow spectral intervals. First, airborne hyperspectral sensors were designed in the early 1980s to extend the scope of remote sensing. Nowadays, they offer opportunities for much more precise identification of forest surface phenomena than is possible with broad-band sensors [33, 34]. Spatial data from each spectral interval is arranged in layers that can be managed in the framework of the image cube with two dimensions formed by the (x) and (y) spatial axis of the image display, and the third (z) formed by the accumulation of spectral data as additional bands, Figure 7. Recently, the hyperspectral data is often captured by airborne instruments such as the Compact Airborne Spectrographic Imager (CASI) [35], Airborne Visible/Infrared Imaging Spectrometer (AVIRIS) [36] and other sensors.

The active remote sensing devices, such as radars, supply their own source of energy to illuminate features of interest. Because active systems do not rely on reflected light from other sources, imagery can be acquired day or night. In the case of forestry and vegetation applications, the radar data is used as a multi-purpose data source that can eliminate imperfection of described techniques in areas of severe cloud cover limitations. Airborne research programs, such as AIRSAR (the Airborne Synthetic Aperture Radar operated by NASA/JPL and flown over many forest sites in the U.S. and Europe) and GLOBESAR (operated by the Canada Centre for Remote Sensing) in recent years have served increase interest in radar in forestry, especially in the estimation of biomass and timber volume [37]. Other active remote sensing devices are represented by Lidarthat has the potential to provide forest canopy height and structure measurements [38].

Figure 7. Hyperspectral data arranged in the image cube with two dimensions formed by the (x) and (y) spatial axis of the image display, and the third (z) formed by the accumulation of spectral data as additional bands.

Remote sensing covers two main activities focused on, as mentioned, data collection by sensors designed to detect electromagnetic energy from sites on Earth surface, and the methods of interpreting the data. After geometric and radiometric corrections, the digital image classification is used to derive information from the numerical representation of images [39]. A few categories of methods can by employed to explore economic, social, and environmental implication of human activities and impacts on forests.

Supervised spectral classifications use information about surface features to aid the computer analysis of their spectral characteristics. The sample sites with known spectral characteristics represent a training set that is used to identify each pixel in the image.

Unsupervised spectral classifications use numerical methods to provide natural groupings within an image. After all pixels in the image are assigned to a defined number of classes, they must be evaluated and named by the analyst.

The spatial pater recognition allows classification of the objects assembled by a set of pixels. The classification is based on spatial relationship with surrounding cells that is defined by a set of rules [40].

Recently, a wide range of other sensors and methods are testing to support the potential role of remote sensing as an information resource to support sustainable forest management. In the case of the spruce forest management, remote sensing is the only way to acquire synoptic and repetitive vegetation observations for large geographic areas over long periods of time.

4. GPS FOR FOREST MANAGEMENT AND PROTECTION

The challenge of determining locations has expanded the frontiers of science with innovative solutions that are represented by Global Positioning Systems (GPS) in this age. It consists of a constellation of satellites orbiting the Earth, several ground stations, and millions of users with field receivers that can provide accurate positions [41, 42]. The U.S. Forest Service has been using GPS to determine actual boundaries of timber sale areas. The field measurements with GPS are more accurate conventional map techniques complemented by remote sensing data. Other applications can include field mapping of the trees attacked by insects in view of local climate conditions and the monocultural type of vegetation. The attacked trees are regularly observed in order to explore migrations of the insects. Nowadays, integration of GPS, GIS and advanced filed computers offers much more than only a collection of locations included into point, line and polygon layers. Field observations in the framework of forest management can be complemented by estimates of tree parameters, calculation of empirical rules and simulation of ecological models [43, 44].

5. MODELING FOR FOREST MANAGEMENT AND PROTECTION

Modeling in forestry can be improved by remote sensing data and field measurements over space and time. The integration of GIS with models in forestry is emerging as a significant new area of the GIS development. Over decades, the model development in the GIS environment has been influenced by a limited number of ways of handling time within the structures provided by a technology that has its roots in the processing of the static contents of maps models [45]. Spatial modeling focused on map algebra, network analysis and 3D visualization are classic GIS applications and are well-suited to this traditional architecture. Recent GISs are able to assist forest managers with dynamic modeling through a process of iteration, in which standard GIS functions are used to transform the system at each step so that the output of one time step becomes the input for the next time step. Thus, a number of dynamic models [46, 47, 48] can be implemented in the GIS in spite of a few problems that stand in the way of this. They include, first of all, a shortage of efficient functions for iterations and the poor performance of the system caused by scripting languages. In order to resolve these barriers, coupling is often used to link stand-alone dynamic models into GIS, including models developed to simulate particular environmental processes in areas such as forest management and protection [49, 50, 51, 52, 53]. The schema in Figure 8 illustrates data flows in the framework of GIS and dynamic modeling tools. The inputs, such as a set of digital maps, spatial data from remote sensing, GPS observations and historical pieces of information are processed in order to provide data sharing in GIS for spatio-temporal analysis of forest succession. GISs support a wide range of spatial formats and data structures that can be imported to simulation tools for the processing of dynamic models. This coupling is most successful with models that predict outcomes of processes such as succession and regeneration in forest ecosystems. Spatial distribution, coincidence, or proximity of observations identified with the GIS and data from supporting tools are input to dynamic models to examine hypothesized predictions in space and time. The simulation outputs can be re-entered into the GIS to design maps of the predicted spruce ecosystem properties.

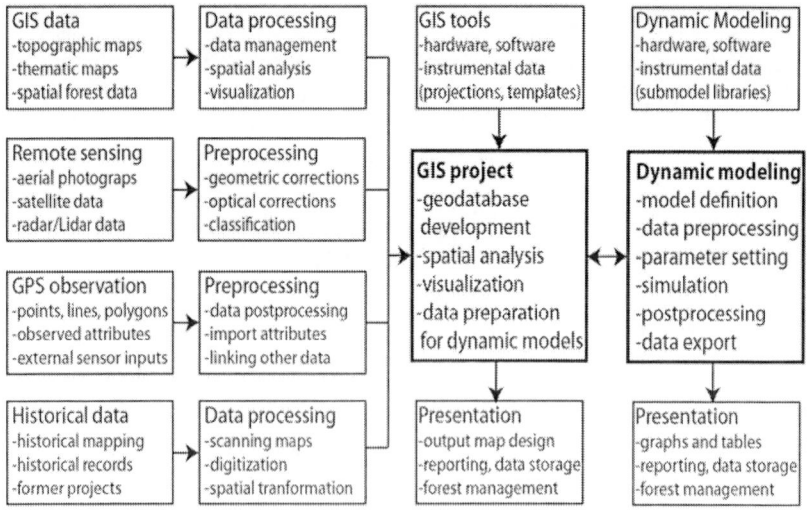

Figure 8. Data processing and modeling in the framework of GIS and dynamic modeling.

6. A CASE STUDY: USING GIS AND DYNAMIC MODELING TO STUDY THE SUCCESSION OF A NORWAY SPRUCE FOREST

Spatio-temporal modeling in forestry based on a wider range of information from remote sensing, GPS-supported observations, and GIS spatial analysis can significantly improve ecological research, management and conservation of spruce forests. The presented case study is focused on spatio-temporal modeling of ground vegetation development in mountain spruce forest. The area of interest is explored by field measurements and remote sensing techniques, Figure 9. Data about trees and attributes of the randomly located experimental research plots with explored seedlings and ground vegetation is managed by GIS, Figure 10. In order to study time series of local spruce survival, the area of interest is divided into sub-areas, microsites. Each local spruce survival is described by an ordinary differential equation that includes a parameter setting of interactions between the local spruce survival and its microsite environmental conditions. Thus, the spatio-temporal model includes a set of ordinary differential equations for the whole area of interest. Many local factors such as the canopy gaps based on estimates of the local crown projected area, the soil type layer, and the dominant grass density can be included to provide case simulation studies. The numerical solution has to be carried out in a stand-alone simulation package for dynamic modeling and simulation. Output data is represented by time series and final state of the spruce forest sites after the selected simulation period. The surviving trees and number of seedlings in microsites are imported in the GIS as new layers. It offers to study spatio-temporal characteristics of the area of interest in the spruce forest in a more complex way together with other data.

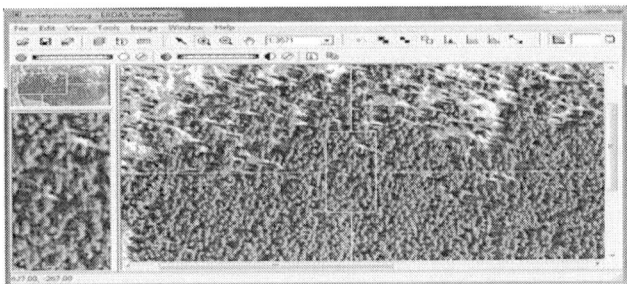

Figure 9. Mapping an area of interest with the aerial photograph (original spatial resolution 0.2 meter).

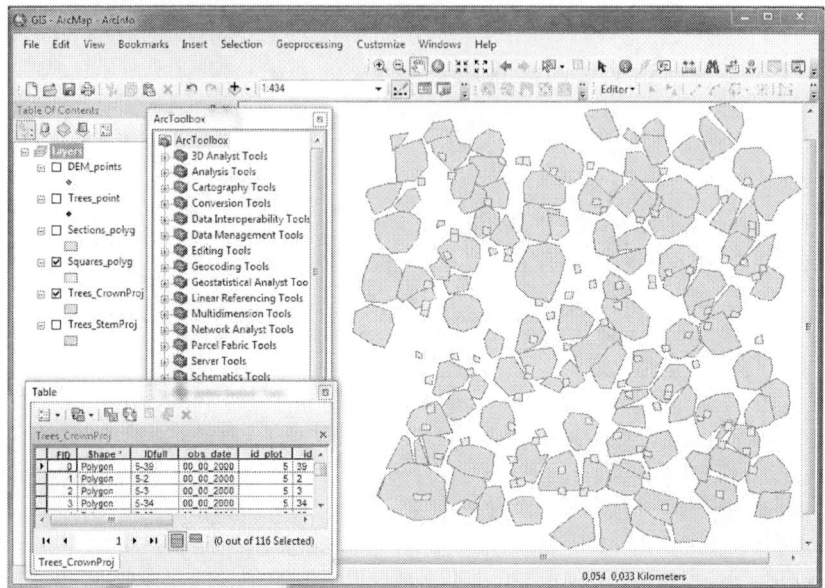

Figure 10. Spatio-temporal data management for exploration of the Norway spruce forest succession (GIS layers show the tree crown projections, and experimental research plots of ground vegetation).

The survival of seedling in the microsite located by coordinates x and y focused on i^{th} age group linked to the defined seed year can be described with an ordinary differential equation:

$$\frac{dN_{x,y,i}}{dt} = f(local\ abiotic\ and\ biotic\ interactions). \quad (1)$$

The initial condition is given by the random number of seeds for each microsite linked to the defined seed year:

$$N_{x,y,i}(seed\ year) = f(random\ appearance). \quad (2)$$

The dispersion of the dominant ground vegetation species s_{VSi} such as *Avenella flexuosa*, *Calamagrostis villosa*, *Vaccinium myrtillus* and others can be described by:

$$\frac{\partial s_{VSi}}{\partial t} = D_{VSi}\frac{\partial^2 s_{VSi}}{\partial x^2} + D_{VSi}\frac{\partial^2 s_{VSi}}{\partial y^2} + r_{VSi}s_{VSi}\left(1 - \frac{\sum species\ in\ competition}{K_{VSi}}\right),$$

where D_{VSi} is the coefficient of spatial propagation, r_{VSi} is the growth ratio, and K_{VSi} is the capacity of the microsite for the ith species. The sum of species in competition represents the demands of the local competing species recalculate to the comparable population size of the ith species. The coefficients are dependent on the local environmental conditions. Many case studies can be created in order to utilize various management and conservation rules in spruce forests. The presented spatio-temporal model has to be adapted to the area of interest with the help of field observations, remote sensing techniques and historical information.

CONCLUSION

Presented research focused on the management and conservation of spruce forests highlights the use of GIS, remote sensing, and GPS as valuable data sources for dynamic modeling. More detailed spatio-temporal models focused on microsites local conditions offer a better view on survival of spruce seedlings in dependence on local abiotic and biotic conditions. Combining dynamic models with new technologies such as GIS, remote sensing complemented by the next generation of sensors with the improved spectral and spatial resolution, GPS, and other advanced field measurements can bring new ways of exploration of spruce forests in order to improve their management and conservation.

ACKNOWLEDGEMENT

The presented research was carried out in the framework of the GIS Laboratory research project supported by the Ministry of Education, Youth and Sports. The partial results were presented at the GIS RESEARCH UK, 13th Annual Conference. The software tools ArcGIS and ERDAS Imagine were used for the processing of presented images.

REFERENCES

[1] Franklin, S.E. (2001). *Remote Sensing for Sustainable Forest Management*. London: CRC Press.

[2] Canadian Council of Forest Ministers (1995). *Defining Sustainable Forest Management. A Canadian Approach to Criteria and Indicators.* Ottawa: Canadian Forest Service, Natural Resources.
[3] Forman, R.T.T. (1995). *Land Mosaics: The Ecology of Landscape and Regions.* Cambridge: Cambridge University Press.
[4] Kohm, K.A. and Franklin J.F. (1997). Creating a Forestry for the 21th Century. Washington, DC: Island Press.
[5] Johnston, C.A. (1998). *Geographic Information Systems in Ecology.* London: Blackwell Science.
[6] Matejicek, L., Vavrova and E., Cudlin, P. (2011). Spatio-temporal modeling of ground vegetation development in mountain spruce forest. *Ecological Modelling 222*, 2584-2592.
[7] Burrough, P.A. and McDonnell, A. (1998). *Principles of Geographical Information Systems.* Oxford: Oxford University Press.
[8] Zeiler, M. (1999). *Modeling Our World.* Redlands, California: ESRI Press.
[9] Arctur, D. and Zeiler, M. (2004). *Designing Geodatabases: Case Studies in GIS Data Modeling.* Redlands, California: ESRI Press.
[10] MacDonald, A. (2001). *Building a Geodatabase.* Redlands, California: ESRI Press.
[11] Maguire, D.J., Batty, M. and Goodchild, M.F. (2005). *GIS, Spatial Analysis and Modeling.* Redlands, California: ESRI Press.
[12] Mitchell, A. (2005). *The ESRI Guide to GIS Analysis. Volume 2: Spatial Measurements & Statistics.* Redlands, California: ESRI Press.
[13] Boyd, D.S. and Foody, G.M. (2011). An overview of recent remote sensing and GIS-based research in ecological informatics. *Ecological Informatics 6*, 25-36.
[14] Campbell, J.B. (2002). *Introduction to Remote Sensing.* 3rd edition. New York: Taylor & Francis.
[15] Landgrebe, D.A. (2003). *Signal Theory Methods in Multi-spectral Remote Sensing.* New Jersey: John Wiley and Sons.
[16] MacLean, D.A., MacKinnon, W.E., Porter, K.B., Beaton, K.P., Cormier, G. and Morehouse, S. (2000). Use of forest inventory and monitoring data in the spruce budworm decision support system. *Computers and Electronics in Agriculture 28*, 101-118.
[17] Wolter, P.T., Townsend, P.A., Sturtevant, B.R. and Kingdon (2008). Remote sensing of the distribution and abundance of host species for spruce budworm in Northern Minnesota and Ontario. *Remote Sensing of Environment 112*, 3971-3982.

[18] Meroni, M., Panigada, C., Rossini, M., Picchi, V., Cogliati, S. and Colombo, R. (2011). Using optical remote sensing techniques to track the development of ozone-induced stress. *Environmental Pollution 157*, 1413-1420.
[19] Shunlin, L. (2004). *Quantitative remote sensing of land surfaces*. New Jersey: Wiley.
[20] Haara, A. and Nevalainen, S. (2002). Detection of dead or defoliated spruces using digital aerial data. *Forest Ecology and Management 160*, 97-107.
[21] Bütler, R. and Schlaepfer, R. (2004). Spruce snag quantification by coupling color infrared aerial photos and GIS. *Forest Ecology and Management 195*, 325-339.
[22] Tuominen, S. and Pekkarinen, A. (2005). Performance of different spectral and textural aerial photograph features in multi-source forest inventory. *Remote Sensing of Environment 94*, 256-268.
[23] Chen, J., Zhu, X., Vogelmann, J.E., Gao, F. and Jin, S. (2011). A simple and effective method for filling gaps in Landsat ETM+ SLC-off images. *Remote Sensing of Environment 115*, 1053-1064.
[24] Vogelmann, J.E., Tolk, B. and Zhu, Z. (2009). Monitoring forest changes in the southwestern United States using multi-temporal Landsat data. *Remote Sensing of Environment 113*, 1739-1748.
[25] Wimberly, M.C. and Reilly, M.J. (2007). Assessment of fire severity and species diversity in the southern Appalachians using Landsat TM and ETM+ imagery. *Remote Sensing of Environment 108*, 189-197.
[26] Meng, Q., Cieszewski, C. and Madden, M. (2009). Large area forest inventory using Landsat ETM+: A geostatistical approach. *ISPRS Journal of Photogrammetry and Remote Sensing 64*, 27-36.
[27] Van Wagtendonk, J.W., Root, R.R. and Key, C.H. (2004).Comparison of AVIRIS and Landsat ETM+ detection capabilities for burn severity. *Remote Sensing of Environment 92*, 397-408.
[28] Zhang, Q., Pavlic, G., Chen, W., Latifovic, R., Fraser, R. and Cihlar, J. (2004). Deriving stand age distribution in boreal forests using SPOT VEGETATION and NOAA AVHRR imagery. *Remote Sensing of Environment 91*, 405-418.
[29] Broadbent, E.N., Asner, G.P., Claros, M.P., Palace, M. and Soriano, M. (2008). Spatial partitioning of biomass and diversity in a lowland Bolivian forest: Linking field and remote sensing measurements. *Forest Ecology and Management 255*, 2602-2616.

[30] Thenkabail, P.S., Hall, J., Lin, T., Ashton, M.S., Harris, D. and Enclona, E.A. (2008). Detecting floristic structure and pattern across topographic and moisture gradients in a mixed species Central African forest using IKONOS and Landsat-7 ETM+ images. *International Journal of Applied Earth Observation and Geoinformation 4*, 255-270.

[31] Leboeuf, A., Beaudoin, A., Fournier, R.A., Guindon, L., Luther, J.E. and Lambert, M.C. (2007). A shadow fraction method for mapping biomass of northern boreal black spruce forests using QuickBird imagery. *Remote Sensing of Environment 110*, 488-500.

[32] Ozdemir, I. and Karnieli, A. (2011). Predicting forest structural parameters using the image texture derived from WorldView-2 multi-spectral imagery in a dryland forest, Israel. *International Journal of Applied Earth Observation and Geoinformation 13*, 701-710.

[33] Van der Meer, F. (2006). The effectiveness of spectral similarity measures for the analysis of hyperspectral imagery. *International Journal of Applied Earth Observation and Geoinformation 8*, 3-17.

[34] Thenkabail, P.S., Enclona, E.A., Ashton, M.S. and Van der Meer, B. (2004). Accuracy assessment of hyperspectral waveband performance for vegetation analysis applications. *Remote Sensing of Environment 91*, 354-376.

[35] Babey, S.K. and Anger C.D. (1993). Compact airborne spectrographic imager (CASI): a progress review. *SPIE 1937*, 152-163.

[36] Xiao, Q., Ustin, S.L. and McPherson, E.G. (2004). Using AVIRIS data and multiple-masking techniques to map urban forest tree species. *International Journal of Remote Sensing 25*, 5637-5654.

[37] Israelsson, H., Ulander, L.M.H., Askne, J.I.H., Fransson, J.E.S., Frölind, P.O., Gustavsson, A. and Hellsten, H. (1997). Retrieval of Forest Stem Volume Using VHF SAR. *IEEE Transactions on Geoscience and Remote Sensing 35*, 36-40.

[38] Lim, K., Treitz, P., Wulder, M., St-Onge, B. and Flood M. (2003). LiDAR remote sensing of forest structure. *Progress in Physical 27*, 88-106.

[39] Tso, B. and Mather, P.M. (2001). *Classification Methods for Remotely Sensed Data*. London: Taylor & Francis.

[40] Blaschke, T. (2010). Object-based image analysis for remote sensing. *ISPRS Journal of Photogrammetry and Remote Sensing 65*, 2-16.

[41] Grewal, M.S., Weill, L.R. and Andrews, A.P. (2001). *Global Positioning Systems, Inertial Navigation, and Integration.* New York: Wiley.

[42] Trimble (2007). *GPS: The First Global Navigation Satellite System.* Sunnyvale, California: Trimble.
[43] Matejicek, L. (2003). Development of Software Tools for Ecological Field Studies Using ArcPad. http://gis.esri.com/library/userconf/proc03/p0333.pdf
[44] Matejicek, L. (2010). Environmental Modeling with GPS. New York: Nova Science Publishers.
[45] Goodchild, M.F., Steyaert, L.T. and Parks, B.O. (1996). *GIS and Environmental Modeling: Progress and Research Issues.* Fort Collins: GIS World, Inc.
[46] Botkin, D.B. (1993). *Forest Dynamics.* New York: Oxford University Press.
[47] Smith, J.M. (1972). *Models in Ecology.* London: Cambridge University Press.
[48] Okubo, A. (1980). *Diffusion and Ecological Problems.* New York: Springer-Verlag.
[49] Pennanen, J., Greene, D.F., Fortin, M.J. and Messier C. (2004). Spatially explicit simulation of long-term boreal forest landscape dynamics: incorporation quantitative stand attributes. *Ecological Modelling 180,* 195-209.
[50] Skjøth, C.A., Geels, C., Hvidberg, M., Hertel, O., Brandt, J., Frohn, L.M., Hansen, K.M., Hedegaard, G.B., Christensen, J.H. and Moseholm,L. (2008). An inventory of tree species in Europe—An essential data input for air pollution modeling. *Ecological Modelling 217,* 292-304.
[51] Gudrun, W., Tappeiner, U., Strobl, J. and Tasser, E. (2008). Understanding alpine tree line dynamics: An individual-based model. *Ecological Modelling 218,* 235-246.
[52] Sohl, T. and Sayler, K. (2008). Using the FORE-SCE model to project land-cover change in the southeastern United States. *Ecological Modelling 219,* 49-65.
[53] Vacek, S., Bastl, M., Leps, J. (1999). Vegetation changes in forest of the Krkonose Mts. over a period of air pollution stress (1980-1995). *Plant Ecology 143,* 1-11.

In: Spruce
Editors: K. I. Nowak and H. F. Strybel © 2012 Nova Science Publishers, Inc.
ISBN 978-1-61942-494-4

Chapter 5

VITAMIN C AS A STRESS BIOINDICATOR OF NORWAY SPRUCE: A CASE STUDY IN SLOVENIA

Samar Al Sayegh Petkovšek[*1] *and Boštjan Pokorny*[1]
[1]ERICo Velenje, Ecological Research and Industrial Cooperation, Velenje, Slovenia

ABSTRACT

Physiological condition of Norway spruce (Picea abies (L.) Karst.) and consequently vitality of forest ecosystems was intensively studied in the period 1991 – 2007 in the northern Slovenia, i.e. in area, influenced by the Šoštanj Thermal Power Plant (ŠTPP). ŠTPP, which is the largest Slovene thermal power plant, used to be the largest Slovene emission source of gaseous pollutants (e.g. SO2, NOx), and very important source of different inorganic (e.g. heavy metals) as well as organic toxic substances (e.g. PAHs). However, extremely high SO2 emission (up to 86,000 t in 1993, and > 120.000 in 1980's, respectively) and dust emissions (up to 8,000 in 1993), have been dramatically reduced after the installation of desulphurization devices in late 1990's. Indeed in the comparison with 1993, SO2 emissions in 2007 were reduced for more then 15-folds and dust emissions for more then 35-folds, respectively. These extreme exposures in the past as well as huge changes in

---

* Corresponding author. Phone: +386 (0)3 898 1953; fax: +386 (0)3 898 1942; e-mail address: samar.petkovsek@erico.si.

environmental pollution during last two decades have significantly influenced vitality of forest ecosystems including physiological conditions (e.g. contents of antioxidant) of different tree species in the study area. Therefore, vitamin C (ascorbic acid) as a sensitive, non-specific bioindicator of stress caused either by anthropogenic (e.g. air pollution) or natural stressors (climatic conditions, diseases, altitude gradient, etc) was included in a permanent survey of forest condition in northern Slovenia. Atmospheric pollutants such as ozone and sulphur dioxide cause formation of free radicals, which are involved in oxidation of proteins and lipids and injury of plant tissues. Plant cells have evolved a special detoxification defence system to cope with radicals, including formation of water-soluble antioxidant, such as vitamin C. The most significant findings and conclusions of the present study are as follows: (a) Vitamin C is a good bioindicator of oxidative stress and an early-warning tool to detect changes in the metabolism of spruce needles, although we found untypical reaction of antioxidant defence in the case of extremely high SO_2 exposure. (b) Metabolic processes in spruce needles react to air pollution according to severity of pollution and the time of exposure. However, if spruce trees were exposed to high SO_2 ambient levels and/or for a long period of time, the antioxidant defence mechanism would be damaged and the content of vitamin C would not increase as expected. (c) Lower exposure to ambient pollution results in better vitality of trees (e.g. higher contents of total (a + b) chlorophyll), as well as in rising of their defence capabilities (higher contents of vitamin C). (d) Physiological condition of Norway spruce in northern Slovenia has significantly improved since 1995, when the desulphurization devices were built on the ŠTPP, and when emissions of SO_2 as well as heavy metals started dramatically and continuously decreasing in this part of Slovenia.

1. INTRODUCTION

Atmospheric (including pollution) and climate changes along with increasing demands upon the forest resources, are three main factors which affected the forest health status at global and local scale (Percy, 2002). The disturbance in supply and allocation of water, nutrients and energy affected the productivity of forest ecosystems and their resistance to biotic and abiotic stressors (Mc Laughlin & Percy, 1999). As a result of human activities (industrial processes, traffic, use of chemical agents in agriculture and households, combustion of fossil fuels, etc.), plants are exposed to far greater amounts of harmful substances than before; moreover, the situation is even more crucial since the environment has been confronted with totally new substances, and consequently plants are not (as yet) adapted to them (Larcher, 1995; Market *et al.*, 2003). The nature of the damages caused by individual chemical substances are modified by the other environmental and stress-inducing factors, resulting in a multiplying effects which usually exceed the tolerant level of organisms (*ibid.*)

Biochemical and physiological indicators of stress caused atmospheric pollutants are used in many studies considering forest health status (Batič *et al.*, 2001; Simončič, 2001; Nyberg *et al.*, 2001; Fürst *et al.,* 2003; Haberer *et al.*, 2006; Al Sayegh Petkovšek *et al.,* 2008; Hofer *et al.,* 2008). Among them, vitamin C (ascorbic acid) is often used as a very promising, sensitive, but rather non-specific bioindicator of environmental stress.

Vitamin C is one of the most important vitamin in human diet, obtained mainly from vegetables, fruits and other plant material. It is implicated in many physiological processes (for example in photosinthesis by regulation of electron flow); moreover, vitamin C is an essential co-factor for the synthesis of zeaxanthin. However, the most significant is its role of being an antioxidant (Foyer, 1993; Larcher, 1995; Cross *et al.,* 1998; Perl-Trevers & Perl, 2002; Esposito *et al.,* 2009). Atmospheric pollutants (sulphur dioxide, nitrogen oxides, ozone, peroxyacetly nitrat (PAN), hydrocarbons, etc.) cause the formation of free radicals, which are involved in the oxidation of proteins and lipids; moreover, injury of several plant tissues can also appear. Vitamin C is an exceptional antioxidant that scavenges, either directly or indirectly, all of the damaging free radicals commonly encountered in plant cell (Foyer, 1993). As primary antioxidant it reacts with hydrogen peroxide (H_2O_2), with superoxid (O_2^-), hydroxil radical (OH^-), and lipid hydroperoxides (Yu, 1994), respectively; furthermore, it is also important secondary antioxidant since it maintenance the α-tocopherol (vitamin E) pool to cope with radicals in deeper layer of membranes. Vitamin E is an efficient lipophilic antioxidant which is incorporated into photosynthetic membranes, and serves to reduce the possibility of damages caused by singlet oxygen or lipid peroxides (Foyer, 1993; Hofer *et al.*, 2008).

In last two decades, permanent survey (biomonitoring) has been performing in northern Slovenia with the aim to assess the forest health condition in the emission area of the largest Slovenian thermal power plant of Šoštanj (ŠTPP) by using the Norway spruce (*Picea abies* (L.) Karst.) needles as bioindicator. In the present paper, a particular attention is focused on the determination antioxidant defence mechanisms (e.g. content of vitamin C) against the pollution stress in Norway spruce, and on assessing the bioindicative value of vitamin C as a sensitive and early-warning bioindicator of environmental pollution with inorganic substances.

2. MATERIAL AND METHODS

2.1. Study Area and Sampling Procedures

The study area is (used to be) exposed to huge amounts of pollutants due to its close vicinity to the largest Slovene thermal power plant of Šoštanj. It has been emitting huge amounts of SO_2, NO_x and CO_2 (Table 1); moreover, annual emissions of heavy metals reached up to 298 t of Zn, 60.6 t of Cr, 22.1 t of Pb, 4.5 t of As, 0.3 t of Hg, and 0.2 t of Cd before installation of desulphurization devices in 1995 and 2000, respectively.

Table 1. Annual emissions of SO_2, NO_x, CO, CO_2 and dust from ŠTPP in the period 1991-2007 (source: Rotnik, 2008)

year	SO_2 (t)	NO_x (t)	CO (t)	CO_2 (t)	dust (t)
1991	80,390	11,057	440	3,142,725	7,495
1992	79,988	9,009	505	3,587,029	6,085
1993	86,101	9,770	523	3,731,473	8,121
1994	80,516	9,483	484	3,434,461	4,917
1995*	51,663	10,025	761	3,581,956	2,765
1996	51,804	10,154	626	3,287,774	1,845
1997	53,093	11,572	739	3,698,747	2,377
1998	55,053	11,963	734	3,821,570	2,316
1999	47,665	9,096	589	3,334,.732	1,077
2000*	44,253	10,379	541	3,540,040	460
2001	18,071	11,403	693	3,887,053	467
2002	22,871	12,779	931	4,740,476	632
2003	13,334	10,936	1,033	4,366,652	480
2004	7,951	8,877	1,300	4,536,876	419
2005	10,341	9,054	1,236	4,622,632	332
2006	6,190	9,130	1,394	4,662,431	158
2007	5,450	8,600	1,269	4,906,889	262
all together	714,734	173,287	13,798	66,883,516	40,208

Note: *Two desulphurization devices were installed in February 1995 and in November 2000, respectively.

A pronounced impact of air pollution was observed in many environmental segments in the study area (e.g. soils and vegetables: Kugonič & Stropnik, 2001; forest stands: Ribarič Lasnik et al., 2001; Al Sayegh Petkovšek et al., 2008; mushrooms: Al Sayegh Petkovšek et al., 2002; lichens: Poličnik et al., 2004, 2008; aquatic organisms: Mazej & Germ, 2009; wild-living ruminants: Pokorny, 2000, 2006; Pokorny et al., 2004).

ŠTPP is located at the bottom of the Šalek Valley, at an altitude of 370 m, in the central northern part of Slovenia (Figure 1), in the apline and pre-alpine vegetation province with moderate continental climate. Prevailing winds are from the west and east, which has an important impact on the distribution of pollutants in the area. In this respect it is important that the ground layer of the frequent thermal inversions usually does not exceed 100 m, which is far below the height of the power station chimneys. Therefore, pollutants are spread over the hilly margins up to 1100 m above sea level, where the upper inversion layer occurs (Šalej, 1999).

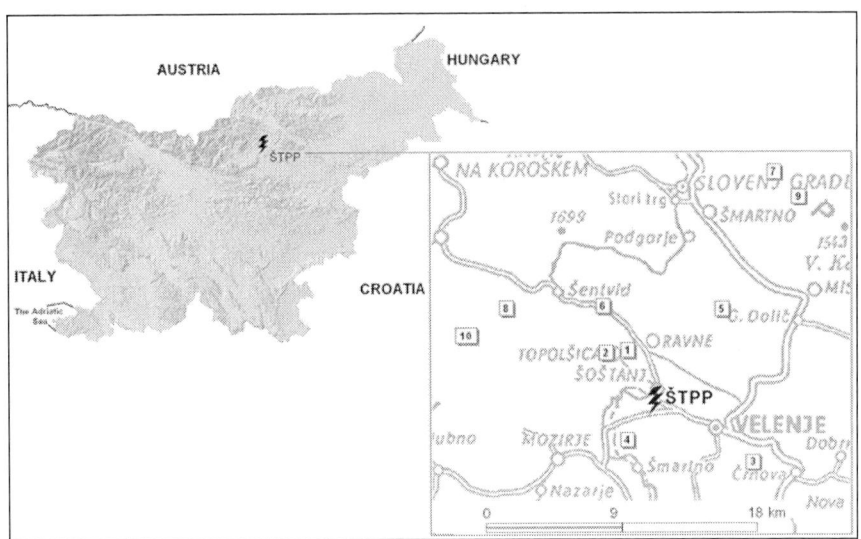

Figure 1. The locations of the Šoštanj Thermal Power Plant (ŠTPP) and sampling sites in the emission area of ŠTTP. The sampling sites are marked with figures, which are shown in Table 2.

To assess the condition of forests in the study area, spruce needles have been used as a bioindicator for last two decades. Norway spruce, the most common Slovenian forest tree species, is a suitable bioindicator of environmental and man-induced stress, used all over Europe, including

Slovenia (Bermadinger-Stabentheiner, 1995; Vidergar Gorjup et al. 2000; Batič et al., 1995, 1999, 2001; Simončič, 2001; Fürst et al., 2003; Modrzynski, 2003; Muzika et al., 2004; Bytnerowicz et al., 2005; Al Sayegh Petkovšek et al., 2008). Spruce needles were sampled in the autumn in every year in the period 1991 – 2007 at ten sampling sites, which differ regarding altitude, direction and distance from the ŠTPP (Table 2). Special attention in the present paper is focused on location Zavodnje, which is strongly affected by air pollution since it is situated just below the belt of the upper thermal inversion, i.e. where the largest deposition of pollutants is expected (Table 3). There selected spruce trees grow in a very close vicinity of the weather station at which concentrations of sulphur dioxide, nitrogen oxides and ozone in the air have been continuously measured during study period; therefore, data on exposure of sampled trees are precisely known.

Table 2. Description of sampling sites

No. of location	Sampling site	distance from the ŠTPP (m)	direction from the ŠTPP	Altitude (m)
1.	Lajše	3,700	SW	400
2.	Topolšica	5,400	NW	400
3.	Laze	5,700	SE	460
4.	Veliki Vrh	3,500	SW	570
5.	Graška gora	7,600	NE	730
6.	Zavodnje	7,600	NW	760
7.	Brneško sedlo	18,100	NE	1,030
8.	Kramarica	12,700	NW	1,070
9.	Kope	17,500	NE	1,400
10.	Smrekovec	14,600	NW	1,555

Table 3. Mean annual deposition of SO_2 ($\mu g/m^3$) at selected locations in the period 1991-2007 (source: Rotnik, 2008)

year	Zavodnje	Veliki Vrh	Topolšica	Graška gora
1991	50	80	40	30
1992	55	76	58	42
1993	47	58	55	47
1994	49	53	34	50
1995*	26	49	20	27
1996	33	57	20	28
1997	42	53	18	36
1998	43	63	20	32
1999	42	72	17	32
2000*	31	56	18	34
2001	20	51	11	15
2002	19	51	14	16
2003	15	45	16	10
2004	8	30	6	6
2005	12	33	5	6
2006	8	20	4	6
2007	7	14	3	5

Note: *Two desulphurization devices were installed in February 1995 and in November 2000, respectively. Bold figures exceeded the permitted levels defined by Slovene and European legislation (20 µg/m3) (Official Gazzete of the Republic of Slovenia, No. 52/2002; Directive 2008/50/EC of the European parliament and the Council of 21 May 2008 on ambient air quality and cleaner air for Europe).

Sampling of current-year needles of Norway spruce was done following the procedure described in the ICP recommendation (Anonymous, 1987). Five vital trees being 60-100 years old were selected per site; from each tree needles from the seventh spindle of branches from the top were collected. Branches were cut off and left overnight in the dark at 4°C for further processing of needles. Needles were frozen in liquid nitrogen and lyophilized prior to biochemical analysis.

2.2. Biochemical Analysis

Contents of vitamin C, vitamin E and photosynthetic pigments (chlorophyll a and b) were determined by high performance liquid chromatography (Hewlet Packard, 1050) according to Grill and Esterbauer (1973), Bui-Nguyen (1980), Wimanlasiri & Wills (1983) and Pfeifhofer (1989), and total sulphur by colorimetric titration using a AOK-S (adsorbed organic halogens) analyser (Euroglas Research, 1998). Following standard reference materials were used for analytical quality control: vitamin C and vitamin E: Bucks Fluk, no. 95210; chlorophyll a: FLUKA 25739; and chlorophyll b: FLUKA 25749, respectively.

2.3. Statistical Analysis

All results presented in the paper represent annual mean values, calculated on the basis of data provided for selected five spruce trees per sampling site, sampled at ten locations in the emission area of the ŠTPP in the period 1991-2007. *Microsoft Excel* was used for calculation of mean values and standard errors. Existence of correlations between different parameters was tested by calculating *Spearman rank coefficient (R)* using Statistica for Windows 7.1 software package (StatSoft, 2006); the limit of statistical significance was set up at $p < 0.05$. In the following sections, all results are given as mg/g on a dry weight basis.

3. RESULTS

3.1. Mean Annual Contents of Vitamin C in Spruce Needles

Physiological condition of spruce trees, sampled in the area influenced by the ŠTPP, was investigated by determination of contents of vitamin C and photosynthetic pigments in current-year Norway spruce needles. Mean annual concentrations of vitamin C are presented in Figure 2. Since the content of vitamin C in green parts of higher plants had already been confirmed as a suitable bioindicator of oxidative stress caused by air pollution (Perl-Treves & Perl, 2002; Langebartels et al., 2002; Eposito et al., 2009), a determination of relation between mean annual contents of vitamin C in needles and mean annual emissions of sulphur dioxide from ŠTPP was emphasized in this study.

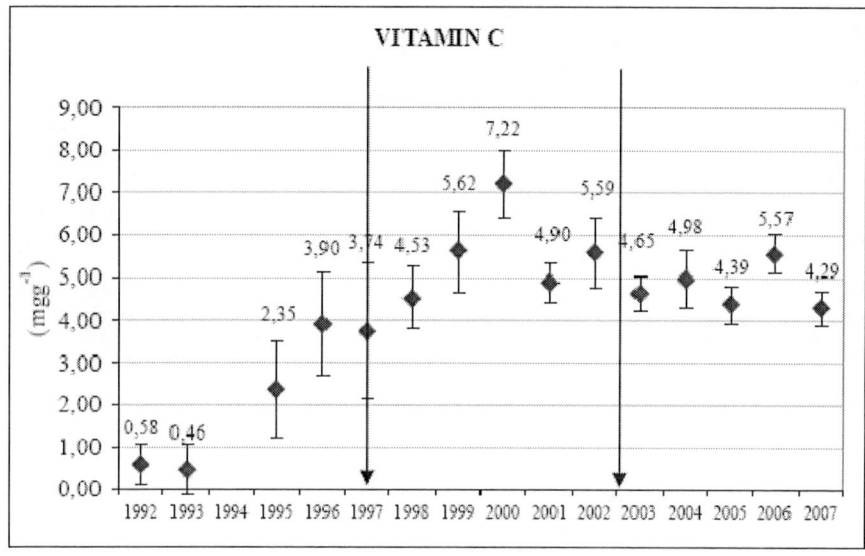

Figure 2. Mean content of vitamin C in current-year needles of spruce located in the emission area of the ŠTPP in the period 1992-2007. The arrows marked the years in which desulphurization devices were installed (Unit 4 in February 1995 and Unit 5 in November 2000, respectively).

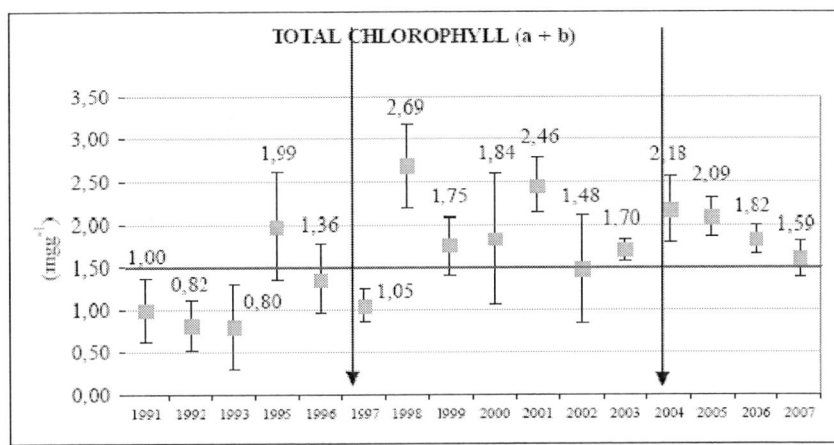

Figure 3. Mean annual concentration of total chlorophyll (a + b) in current-year needles of spruce located in the emission area of ŠTPP. The horizontal line represents the limit value (1.5 mg/g). The arrows marked the years in which desulphurization devices were installed (Unit 4 in February 1995 and Unit 5 in November 2000, respectively).

As a rule, the defence mechanism of plants and consequently content of vitamin C in their tissues should increase with increasing air pollution (Foyer, 1993; Larcher, 1995; Cross *et al.,* 1998; Perl-Trevers and Perl, 2002). However, in the period of the largest emissions of sulphur dioxide (1991-1994), the lowest mean concentrations of vitamin C in spruce needles were found (Figure 2). Immediately after the first significant reduction of SO_2 emissions in 1995, contents of vitamin C in needles started increasing and reaching the peak in 2000, although emissions remained almost unchanged in that period. Such a trend is comparable with some previous studies from highly polluted areas (e.g. Grill *et al.*, 1979; Bermadinger *et al.*, 1990; Batič *et al.*, 2001). If spruces trees were exposed to high SO_2 emissions and for a long time, the antioxidant defence mechanism would be damaged and the content of vitamin C would not increase as expected. In our study area, previous huge emission of SO_2 were firstly significantly reduced after the installation of the desulphurization devices on the fourth unit of the ŠTPP in February 1995, it is most likely that after a long lasting stress in 1980's and 1990's the defence mechanism in spruce needles has started repairing and the normal mechanism of formation of antioxidant has been re-establishing after the implantation of this mitigation measure. After the second significantly reduction of emissions of SO_2 (after the installation of the desulphurization devices on the fifth unit of the ŠTPP in November 2000) the contents of vitamin C drastically diminished and remained almost equal.

In order to assess the health status of investigated trees we also measured the photosynthetic pigments, since oxidative stress tends to reduce chlorophyll content, especially chlorophyll a. The mean annual concentrations of total chlorophyll (a + b) in current-year needles are shown in Figure 3. A trend of increasing spruce vitality after 1995 was confirmed; moreover, the chlorophyll content exceeded the limit value (1.5 mg/g – value indicating tree injury) (Köstner et al., 1990) for the first time in that year and remained above this value until 2007 (with slightly decrease in year 1996, 1997 and 2002). Presumably, decrease of the total pigment contents in these three years were not correlated with air pollution, since emissions of air pollutants remained unchanged in that period. A lower vitality of trees (i.e. lower contents of pigment) in these three years reflects the impact of natural stressors (e.g. high summer T and drought in year 1996 and 2002, respectively).

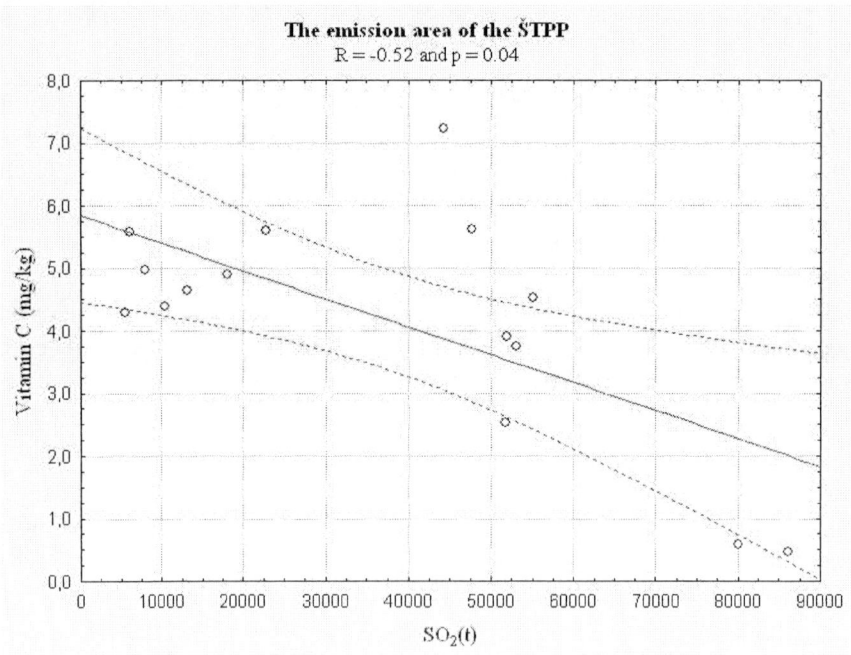

Figure 4. Correlation between mean annual emissions of SO_2 and mean annual contents of vitamin C in spruce needles.

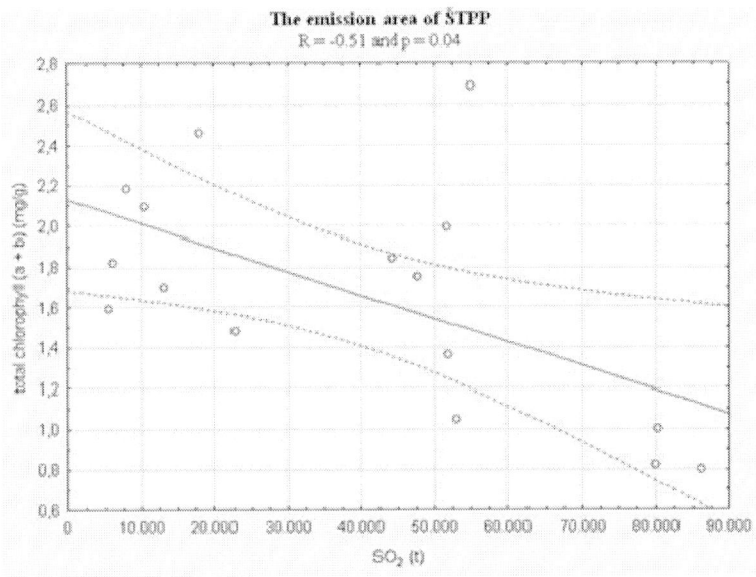

Figure 5. Correlation between mean annual emissions of SO2 and mean annual concentration of total chlorophyll (a + b) in spruce needles.

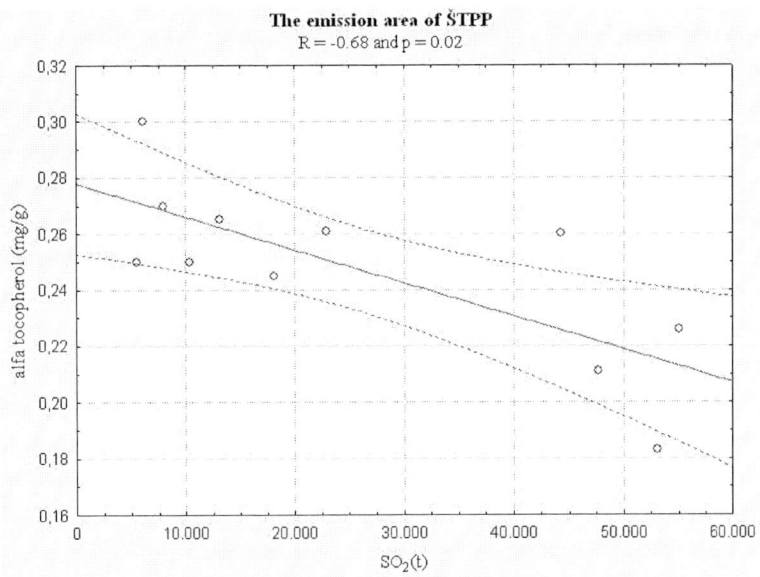

Figure 6. Correlation between mean annual emissions of SO2 and mean annual concentration α tocopherol (vitamin E) in spruce needles.

3.2. Correlation Analysis

Since content of vitamin C in spruce needles, may be affected either by natural stressors (extreme temperatures, drought, ultraviolet radiation, etc.) or by air pollution, we tested the existence of correlation between two stress-inducing factors (altitude and deposition of SO_2 at selected locations) and mean annual content of vitamin C in spruce needles. Due to untypical reaction of trees to extremely high exposure to SO_2 in the first study years we confirmed the existence of negative correlation between mean annual SO_2 emissions and mean annual concentration of ascorbic acid for the study period (Figure 4). Moreover, mean annual concentrations of total chlorophylle (a + b) and vitamin E were significantly affected by SO_2 emissons (Figures 5, 6; Table 4), as well.

Table 4. Correlations between mean annual concentrations of selected parameters, in spruce needles in the emission area of ŠTPP

	vitamin E	chlorophyll (a + b)	emission of SO_2
vitamin C	ns	ns	ns
α tocopherol	-	ns	R = -0.68; p = 0.02
chlorophyll (a + b)	-	-	R = -0.51; p = 0.04

Note: ns: not significant.

Table 5. Correlations between annual concentrations of selected parameters, in spruce needles for location Zavodnje

	vitamin E	chlorophyll (a + b)	deposition of SO_2	deposition of O_3
vitamin C	ns	R = 0.57; p = 0.03	ns	R = -0.52; p = 0.06
α tocopherol	-	ns	ns	ns
chlorophyll (a + b)	-	-	R = -0.56; p = 0.03	ns
deposition of SO_2	-	-	-	ns

Note: ns: not significant

The impact of altitude on vitamin C content in spruce needles was tested per every single year. It was established that in study area altitude does not affected the content of vitamin C in spruce needles. Normally (e.g. in areas, where the impacts of air pollution is absent), the content of vitamin C increase with rising altitude. For northern Slovenia, however, it is evident that stress caused by higher altitude does not prevail over the stress caused by air pollution. Indeed, the locations at lower altitudes are more exposed to the emissions from the ŠTPP, that is indicated by mean annual contents of total sulphur in spruce needles (Table 6).

Table 6. The mean contents of total sulphur (mg/g) in current-year spruce needles

	Lajše	Topolšica	Laze	Veliki Vrh	Graška gora	Zavodnje	Brneško sedlo	Kramarica	Kope	Smrekovec
1991	1.84	1.89	1.42	2.13	1.45	1.57	1.33	1.62	1.26	1.39
1992	2.01	2.14	1.39	2.47	1.57	1.61	1.40	1.88	1.35	1.45
1993	1.99	1.81	1.53	1.51	2.19	1.69	1.57	1.69	1.40	1.42
1994	/	/	/	2.50	/	1.60	1.30	/	/	1.50
1995	1.51	1.60	1,42	1.69	/	1.52	1.00	1,32	0.95	1.11
1996	1.54	1.57	1.24	1.61	/	1.66	1.23	1.50	1.07	1.34
1997	1.66	1.40	1.14	1.69	1.39	1.45	1.31	1.47	1.06	1.07
1998	1.65	1.52	1.40	1.65	1.32	1.35	1.21	1.49	0.99	1.12
1999	1.74	1.59	0.97	1.94	0.37	1.50	1.20	1.35	1.12	1.29
2000	1.75	1.60	1.16	2.19	1.48	1.51	1.31	1.48	1.11	1.12
2001	1.32	1.44	0.97	1.52	1.07	1.32	1.06	1.25	1.09	1.05
2002	1.46	1.46	1.00	1.74	1.18	1.44	1.02	1.07	1.01	1.09
2003	1.24	1.24	0.78	1.22	1.04	1.20	0.96	1.16	0.96	1.08
2004	1.26	1.30	0.96	1.20	1.26	1.10	0.86	0.94	0.90	1.02
2005	0.97	0.97	0.83	1.10	0.97	1.07	0.80	0.95	1.03	0.83
2006	0.97	1.00	0.83	0.87	0.80	0.90	0.70	0.77	0.97	0.90
2007	0.83	0.87	0.83	0.87	0.93	1.03	0.73	0.80	0.80	0.83

Note: Bold figures exceeded levels of total sulphur characteristic for spruce needles from areas which are not loaded with sulphur dioxide (0.97 /kg) (Kalan et al. 1995).

Total sulphur in spruce needles could reflect the exposure of spruce trees to SO_2 emissions, since mean annual total sulphur content in needles was directly correlated with the mean annual SO_2 emissions in the period of 1991–2007 (R = 0.91; p = 0.000001; n = 16). Analyses of single year needles indicated that sulphur content in spruce needles is highest at sites close to the power plant (Veliki Vrh, Topolšica, Lajše) and where altitude coincides with that of frequent thermic inversions (Zavodnje) (Ribarič Lasnik *et al.*, 2001; Tausz *et al.*, 2002).

In addition, ozone concentrations in the air influence the content of vitamin C in spruce needles as well (Larcher, 1995). Since data on ozone deposition are precisely known only for location Zavodnje, the existence of correlation was tested for this sampling plot. There content of vitamin C in spruce needles, was significantly affected by ozone concentrations; indeed, content of vitamin C decreased with increasing concentration of ozone (Figure 8) (R = -0.52; p = 0.056). Due to long lasting influence of air pollution on the physiological status of spruce trees (Table 3), their defence mechanism is damaged; consequently, the elevated concentrations of ozone additionally lead to decrease in the content of vitamin C.

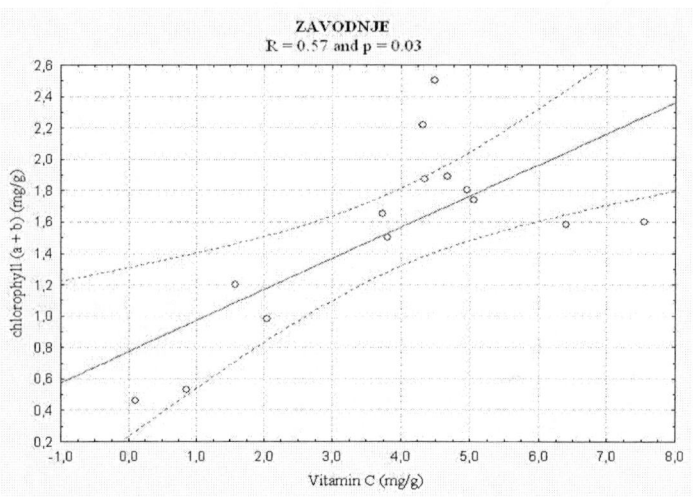

Figure 7. Relation between contents of vitamin C and total chlorophyll (a + b) in spruce needles, sampled at location Zavodnje across the study period.

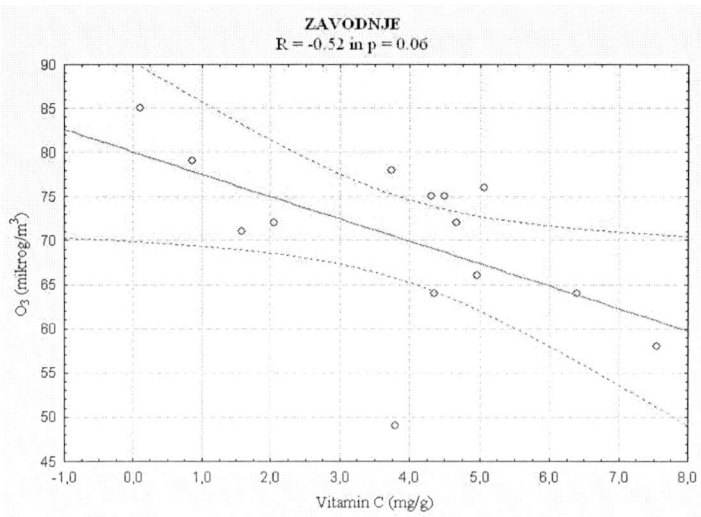

Figure 8. Relation between contents of vitamin C in spruce needles and mean annual imissions of ozone at location Zavodnje across the study period.

Together with temporal increase in annual contents of vitamin C in spruce needles, sampled at Zavodnje, concentrations of photosynthetic pigment increased as well ($R = 0.57$; $p = 0.03$) (Figure 5). It is evident that better physiological status of spruce is correlated with increasing antioxidant production (rising defence capabilities) and better vitality of spruce trees, grown at sampling site Zavodnje.

CONCLUSION

On the basis of data collected during the permanent biomonitoring of forest ecosystem performed in the period 1991-2007 in northern Slovenia the most significant findings and conclusions are as follows: (a) Vitamin C is a good bioindicator of oxidative stress and an early-warning tool to detect changes in the metabolism of spruce needles, although we found untypical reaction of antioxidant defence in the case of extremely high SO_2 exposure. (b) Metabolic processes in spruce needles react to air pollution according to severity of pollution and the time of exposure. However, if spruce trees were exposed to high SO_2 ambient levels and/or for a long period of time, the antioxidant defence mechanism would be damaged and the content of vitamin C would not increase as expected. (c) Lower exposure to ambient pollution results in better vitality of trees (e.g. higher contents of total (a + b) chlorophyll), as well as in rising of their defence capabilities (higher contents of vitamin C). (d) Physiological condition of Norway spruce in northern Slovenia has significantly improved since 1995, when the desulphurization devices were built on the ŠTPP, and when emissions of SO_2 as well as heavy metals started dramatically and continuously decreasing in this part of Slovenia.

ACKNOWLEDGEMENT

Reviewed by dr. Primož Simončič, Slovenian Forestry Institute (Gozdarski inštitut Slovenije), Večna pot 2, 1000 Ljubljana, Slovenija, primoz.simoncic@gozdis.si.

REFERENCES

[1] Al Sayegh Petkovšek, S., Batič, F. & Ribarič Lasnik, C. (2008). Norway spruce needles as bioindicator of air pollution in the area influence of the Šoštanj thermal power plant, Slovenia. *Environmental pollution, 151,* 287-291.
[2] Al Sayegh Petkovšek, S., Pokorny, B., Ribarič Lasnik, C., Vrtačnik, J. (2002). Vsebnost Cd, Pb, Hg in As v trosnjakih gliv iz gozdnate krajine Šaleške doline. *Zbornik gozdarstva in lesarstva, 67,* 5-46.
[3] Anonymous (1987). Manual and methodologies and criteria for harmonized sampling, assessment, monitoring and analyses of the effect of air pollution on forests. *ECE/NEP for Europe.* Geneva.
[4] Batič, F., Kalan, P., Kraigher, H., Šircelj, H., Simončič, P., Vidergar-Gorjup, N. & Turk, B., (1999). Bioindication of different stresses in forest decline studies in Slovenia. *Water Air and Soil Pollution, 116,* 377-382.
[5] Batič, F., Vidergar-Gorjup, N., Šircelj, H., Simončič, P. & Turk, B. (2001). The analyses of photosynthetic pigments, ascorbic acid and macronutrients – a tool for evaluation of the effect of air pollution in Norway spruce (*Picea abies* (L.) Karst.). *Journal of Forest Science 47,* 39-48.
[6] Bermadinger, E., Guttenberger, H. & Grill, D. (1990). Physiology of young Norway spruce. *Environmental Pollution, 68,* 319-330.
[7] Bui-Nguyen, H. M. (1980). Application of high-performance liquid chromatography to the separation of ascorbic acid to the isoascorbic acid. *Journal of Chromatography, 196,* 163-165.
[8] Burmandiger-Stabentheiner, E. (1995). Stress-Physiological Investigation on Spruce Trees (*Picea abies* (L.) Karst.) from the "Achenkirch Altitude Profiles". *Phyton, 29,* 255-301.
[9] Bytnerowitz, A., Badea, O., Popescu, F., Musselman, R., Tanase, M., Barbu, I., Fraczek, W., Gembasu, N., Surdu, A., Danescu, F., Postelnicu, D., Cenusa, R. & Vasile, C. (2005). Air pollution, precipitation chemistry and forest health in the Retezat Mountains, Southern Carpathians, Romania. *Environmental Pollution, 137,* 546-567.
[10] Cross, C.E., van der Vliet, A., Louie, S., Thiele J.J. & Halliwell, B. (1998). Oxidative stress and antioxidants at biosurfaces: plants, skin, and respiratory tract surface. *Environmental Health Perspectives, 106,* 1241-1251.

[11] Directive 2008/50/EC of the European parliament and the Council of 21 May 2008 on ambient air quality and cleaner air for Europe.
[12] Eposito, M. P., Ferreira, M. L., Sant'anna, S. M. R., Domingos, M. & Souza S. R. (2009). Relationship between leaf antioxidant and ozone injury in *Nicotiana tabacum* "Bel-W3" under environmental conditions in São Paulo, SE – Brazil. *Atmosperic Environment, 43,* 619-623.
[13] Euroglass Research & Aplication laboratory, 1998. *Determination of total sulphur in spruce needle samples by ECS 1600/3000, analyses report.*
[14] Föyer, C. H. Ascorbic acid. In: Alscher, R.G., Hess, J. L., editor. *Antioxidants in higher plants.* CRP Press, Inc., 1993, 32-59.
[15] Fränzle, O. (2003). Bioindicators and environmental stress assessment: In: B. A. Market, A. M Breure, H. G. Zechmeister, editors. *Trace Metals and other Contaminants in the Environment 6: Bioindicator & Biomonitors, Principles, concepts and applications,* Elsevier, 41-85.
[16] Fürst, A., Smidt, S. & Herman, F. (2003). Monitoring the impact of sulphur with the Austrian Bioindicator Grid. *Environmental Pollution, 125,* 13-19.
[17] Grill, D. & Esterbauer, H. (1973). Quantitative Bestimung wasserloslicher Sulfhydrylverbindung in gesunden und SO_2-geschädigten Nadeln von *Picea abies. Phyton, 15,* 87-101.
[18] Grill, D., Esterbauer, H. & Welt, R. (1979). Einfluss von SO_2 auf das Ascorbinsäuresystem der Fichtenndeln. *Phytopathology, 96,* 361-368.
[19] Haberer, K., Jaeger, L. & Rennenberg, H. (2006). Seasonal patterns of acorbate in the needles of Scots Pine (*Pinus sylvestris* L.) trees: Correlation analyses with atmospheric O_3 and NO_2 gas mixing ratios and meteorological parameters. *Environmental Pollution, 139,* 224-231.
[20] Hofer, N, Alexou, M., Heerdt, C., Löw, M., Werner, M., Matyssek, R., Rennenberg, H. & Haberer, K. (2008). Seasonal differences and within-canopy variations of antioxidants in mature spruce (*Picea abies*) trees under elavated ozone in a free-air exposure system. *Environmental pollution, 154,* 241-253.
[21] Kalan, J., Kalan, P. & Simončič, P. (1995). Preučevanje gozdnih tal na stalnih raziskovalnih ploskvah. *Zbornik gozdarstva in lesarstva,* 47, 57-84.
[22] Köstner, B., Czygan, F.C. & Lange, O.L. (1990). An analysis of needle yellowing in the healty and chlorotic Norway spruce (*Picea abies*) in a forest decline area of the Fichtelgebirge (NE: Bavaria). *Trees, 4,* 55-67.

[23] Kugonič, N. & Stropnik, M. (2001). Vsebnost težkih kovin v tleh in rastlinah na kmetijskih površinah v Šaleški dolini, *letno poročilo, DP 24/02/01*. Velenje, ERICo, 183 str.
[24] Kunert, K.J. & Ederer, M. (1985). Leaf aging and lipid peroxidation: The role of antioxidant vitamin C and E. *Physiologia Plantarum, 65*, 85-88.
[25] Langebartels, C., Schraudner, M., Heller, W., Ernst, D. & Sandermann, H. (2002). Oxidative stress and defense reaction in plants exposed to air pollution and UV-B radiation. In: D. Inze, M. Van Montagu, editors. *Oxidative stress in Plants*. New York, Taylor & Francis Inc, 105-136.
[26] Larcher, W. *Physiological plant ecology. Ecophysiology and stress physiology of functional groups.* Third edition. Berlin, Heidelberg, Springer Verlag, 1995.
[27] Market B. A., Breure A. M. & Zechmeister H. G. (2003). Definitions, strategies and principles for bioindication/biomonitoring of the environment. In: B. A. Merket, A. M Breure, H. G. Zechmeister, editors. *Trace Metals and other Contaminants in the Environment 6: Bioindicator & Biomonitors, Principles, concepts and applications*, Elsevier Science, 3-39.
[28] Mazej, Z. & Germ, Z. (2009). Trace element accumulation and distribution in four aquatic macrophytes. *Chemosphere, 74*, 642-647.
[29] McLaughlin S.B. & Percy K.E. (1999). Forest health in North America: Some perspective and potential roles of climate and air pollution. *Water air and soil pollution, 116*, 151-197
[30] Modrzynski, J. (2003). Defoliation of older Norway spruce (*Picea abies* (L.) Karst.) stands in the polish Sudety and Carpatian mountains. *Forest Ecology and Management, 181*, 289-299.
[31] Muzika, R. M., Guyette, R. P., Zielonka, T. & Liebhold A. M. (2004). The influence of O_3, NO_2 and SO_2 on growth of *Picea abies* and *Fagus sylvatica* in the Carpathian Mountains. *Environmental Pollution, 130*, 65-71.
[32] *Official Gazzete of the Republic of Slovenia*, No. 52/2002.
[33] Percy, K.E. (2002). Is air pollution an important factor in forest health? In: R.C Szaro, editor. *Effects of air pollution on forest health and biodiversity in forests of the Carpatian mountains*. IOS Press, 23-43
[34] Perl-Treves, R. & Perl, A. (2002). Oxidative Stress: An introduction. In: D. Inze, M. Van Montagu, editors. *Oxidative stress in Plants*. New York, Taylor & Francis Inc, 1-32.

[35] Pfeifhofer, H. W. (1989). On the pigment content of Norway spruce needles infected with *Chrysomyxa rhodendendri*, and the carotenoids of the fungus Aeciospores. *European Journal of Forest Pathology, 19*, 363-369.
[36] Poličnik, H., Batič, F. & Ribarič Lasnik, C. (2004). Monitoring of short-term heavy metal deposition by accumulation in epiphytic lichens (*Hypogymnia physodes* (L.) Nyl.). *Journal of atmospheric chemistry, 49*, 223-230.
[37] Poličnik, H., Simončič, P. & Batič, F. (2008). Monitoring air quality with lichens: A comparison between mapping in forest sites and in open areas. *Environmental pollution, 151*, 395-400.
[38] Pokorny, B. (2000). Roe deer *Capreolus capreolus* as an accumulative bioindicator of heavy metals in Slovenia. *Web Ecology, 1*, 54-62.
[39] Pokorny, B. (2006). Roe deer (*Capreolus capreolus* L.) antlers as an accumulative and reactive bioindicator of lead pollution near the largest Slovene thermal power plant. *Veterinarski Arhiv, 76*, S131-S142.
[40] Pokorny, B., Glinšek, A. & Ribarič Lasnik, C. (2004). Roe deer antlers as a historical bioindicator of lead pollution in the Šalek Valley, Slovenia. *Journal of atmospheric chemistry, 49*, 175-189.
[41] Ribarič Lasnik, C., Bienelli Kalpič, A., Batič, F. & Vrtačnik, J. (2001). Biomonitoring of forest ecosystem in areas influenced by the Šoštanj and Trbovlje Thermal Power Plants. *Journal of Forest Science, 47*, 61-67.
[42] Rotnik, U. Bilten TEŠ. *Poročilo o proizvodnji, vzdrževanju in ekoloških obremenitvah okolja TE Šoštanj v letu 2007.* Velenje, AV Studio d.o.o, Velenje, 2008.
[43] Šalej, M. Historično-geografski oris Šaleške doline in njenega obrobja. In: T. Ravnikar, editor. *Razprave o zgodovini mesta in okolice.* Velenje, Mestna občina, 11-37.
[44] Simončič, P. (2001). Nutrient condition for spruce (*Picea abies* (L.) Karst.) in the area affected by a Thermal Power Plant in Slovenia. *Journal of Forest Science, 47*, 67-72.
[45] Šircelj, H., Batič, F. & Štampar, F. (1999). Effects of drought stress on pigment, ascorbic acid and free acid content in leaves of two apple tree cultivars. *Phyton, 39*, 97-100.
[46] StatSoft 2006. *Statistica for Windows 7.1.* Tulsa, StatSoft, CD.
[47] Stefan, K. (1990). Vergleich der schwefelanalysedated des Österreichen Bioindikatornetzes in Jahr 1989 mit den Etgebnissen vorangegangener

Jahre. (Bericht BIN-S 62/1990). *Forstliche Bundesversuchsanstalt.* Institut fülmmissionsforschung and Forrstchemie, Wiein, 70 p.
[48] Tausz, M., Morales, D., Jimenez, M.S. & Grill, D. (1999). Photoprotection in forest trees under field condition. *Phyton, 39,* 25-28.
[49] Tausz, M., Wonisch, A., Ribarič Lasnik, C., Batič, F. & Grill, D. (2002). Multivariate Analyses of Tree Physiological Attributes – Applications in Field Studies. *Phyton, 42,* 215-221.
[50] Wimanlasiri, P., Wills, R.B. (1983). Simultaneous analyses of acorbic acid in fruit and vegetables by high performance liquid chromatography. *Journal of Chromatography, 256,* 368-371.
[51] Yu, B. P. (1994). Cellular defenses agains damage from reactive oxgen species. *Physiol. Rev., 74,* 139-162.

INDEX

A

access, 97
age, 4, 8, 11, 15, 16, 18, 19, 22, 34, 105, 108, 111
agencies, 102
agriculture, 116
air pollutants, 124
air quality, 133
alters, 73
ambient air, 121, 131
ammonia, 48, 50, 53, 59, 60, 61, 67, 68, 69, 72
ammonium, 45
amplitude, 88, 90
anaerobic digesters, 50
antioxidant, xi, 116, 117, 123, 129, 131, 132
ascorbic acid, xi, 116, 117, 126, 130, 133
assessment, 112, 130, 131
assimilation, 64, 67
atmosphere, 60, 83, 102

B

bacteria, viii, ix, 41, 42, 47, 48, 49, 51, 53, 54, 55, 59, 60, 61, 63, 65, 66, 69, 70
bacterium, 65
barriers, 106
base, 97
beetles, 6
benefits, 44
biodiversity, 96, 132
biomass, 45, 47, 71, 73, 103, 111, 112
biomonitoring, 117, 129, 132
bioremediation, 73
biotic, 109, 116
biotin, 64
boreal forest, viii, ix, 3, 30, 31, 33, 34, 41, 42, 43, 44, 45, 47, 50, 51, 52, 53, 54, 55, 56, 60, 61, 62, 63, 64, 67, 68, 111, 113
branching, 48
breakdown, 33, 34
browsing, 35
burn, 90, 111

C

calorimetry, x, 76, 80, 89, 90
capillary, 90
carbohydrate, 72
carbon, 44, 45, 47, 48, 49, 51, 60, 61, 62, 63, 64, 66, 67, 68, 73
carotenoids, 133
case studies, 109
case study, x, 39, 96, 107
cellulose, 44
chemical, 44, 116
chimneys, 119
chitin, 45
chlorophyll, xi, 116, 121, 123, 124, 125, 126, 128, 129
chronobiology, x, 76

classes, 5, 6, 8, 9, 11, 12, 14, 15, 16, 19, 20, 22, 31, 32, 33, 104
classification, 35, 67, 97, 104
climate, 3, 4, 31, 38, 105, 116, 119, 132
climate change, 116
clone, 71
clusters, 16, 18, 19, 50, 51, 71
coding, 65
coenzyme, 62
coherence, 84
collaboration, 92
colonisation, 47, 64
colonization, 31, 39, 63
combustibility, 90
combustion, 116
communication, 92
competition, 32, 61, 109
composition, ix, 3, 5, 37, 38, 39, 42, 55, 60, 73
compounds, 31, 47, 49, 62
compression, x, 76, 87, 88
computer, 104
conditioning, vii, 1
configuration, 87
conifer, 37, 38
conservation, vii, x, 96, 98, 99, 107, 109
constituents, 48
contaminated soil, 71
contaminated soils, 71
controversial, x, 76
correlation, 11, 79, 126, 127
correlations, 121
covering, 7, 29, 30
cracks, 31
crop, 43
crown, 107, 108
crowns, 5, 29, 33
cultivars, 133
cultivation, 66
culture, 50, 66
cycles, ix, 76, 79, 91
cycling, 44, 61, 62, 63, 72

D

data analysis, x, 96
data collection, 104
data processing, 97
data set, 78
data structure, 98, 106
database, x, 96, 97, 99
decay, 20, 31, 37, 39, 40, 83, 91
decision-making process, x, 96
decomposition, viii, 2, 5, 6, 8, 9, 11, 12, 13, 14, 15, 17, 19, 20, 22, 24, 25, 29, 31, 32, 33, 34, 71
defence, xi, 116, 117, 123, 128, 129
deforestation, 65
degradation, 44, 49
degradation process, 44, 49
dependent variable, 78
deposition, 29, 30, 32, 34, 120, 126, 127, 133
depth, 4, 71
derivatives, 62
detectable, ix, 42, 58, 59, 63
detection, 53, 55, 58, 61, 63, 64, 103, 111
detoxification, xi, 116
deviation, 90
diet, 117
diseases, xi, 116
dispersion, 108
distribution, 8, 9, 11, 18, 19, 20, 21, 22, 25, 32, 52, 56, 99, 106, 110, 111, 119, 132
diversity, 35, 48, 49, 55, 58, 59, 63, 64, 65, 66, 67, 70, 71, 72, 111
dominance, 3
drawing, 91
drought, 124, 126, 133
drying, ix, 76, 77, 79, 82, 83, 90, 93, 94
durability, 91

E

ecology, vii, x, 36, 39, 50, 63, 96, 99, 132
ecosystem, 73, 96, 106
editors, 36, 37, 38, 39, 131, 132

electric current, 77
electromagnetic, 102, 103, 104
electron, 69, 117
electrophoresis, 65
e-mail, 1, 115
emission, x, 71, 115, 117, 119, 121, 122, 123, 126
endonuclease, 68
energy, 51, 62, 80, 89, 102, 103, 104, 116
engineering, 75, 93
enlargement, 33
erosion, 31
ester, 49
ethylene, 48
Europe, 3, 102, 103, 113, 119, 121, 130, 131
evidence, 64, 77
evolution, 66
exposure, xi, 116, 120, 126, 127, 129, 131
extraction, vii, 2, 63

F

families, 45, 46
fatty acids, 49
fertilization, 73
films, 102
fixation, 44, 51, 53, 61, 68
fluctuations, 77
fluorescence, 65
formation, xi, 3, 48, 69, 116, 117, 123
fragments, 5, 31, 65
free radicals, xi, 116, 117
freshwater, 56
fructose, 48
fruits, 117
funds, 91
fungal metabolite, 47
fungi, ix, 42, 44, 45, 46, 47, 48, 52, 54, 58, 61, 63, 66, 68, 69, 73
fungus, ix, 42, 45, 46, 47, 54, 56, 58, 59, 63, 73, 133

G

gel, 65
genome, 50, 66, 71
genotype, 72
genus, 56, 58, 61, 63
germination, 3, 4, 29, 31, 33, 37, 39, 48, 76, 93
glycerol, 49
glycoproteins, 49
graph, 19
grass, 52, 54, 65, 107
greenhouse, 44
grouping, 50
growth, viii, 2, 3, 27, 28, 29, 30, 31, 33, 34, 36, 37, 38, 39, 40, 48, 51, 60, 62, 67, 76, 86, 87, 109, 132

H

halogens, 121
haze, 102
health, 96, 97, 116, 117, 124, 130, 132
heavy metals, xi, 115, 118, 129, 133
height, viii, 2, 5, 8, 9, 18, 19, 20, 21, 22, 23, 27, 28, 29, 31, 32, 77, 79, 103, 119
height growth, viii, 2
hemisphere, 43
heterogeneity, vii, 1, 4, 28
history, viii, 41, 43, 92
hormones, 47
host, 45, 47, 50, 63, 72, 110
hot springs, 42
human, 96, 104, 116, 117
humus, ix, 5, 36, 42, 44, 46, 48, 53, 54, 55, 56, 58, 62, 63, 64, 67, 72, 73
hybridization, 54
hydrocarbons, 117
hydrogen, 60, 117
hydrogen peroxide, 117
hydroperoxides, 117
hypothesis, 62
hysteresis, 78

I

identification, 65, 103
illumination, 33
image, 95, 98, 99, 100, 101, 102, 103, 104, 109, 111, 112
immersion, 78, 84, 85, 86, 87, 90, 91
improvements, 102
in situ hybridization, 68
individuals, viii, 2, 7, 8, 9, 10, 11, 12, 13, 14, 15, 16, 17, 18, 19, 20, 21, 22, 23, 24, 25, 26, 28, 31, 34
industry, 96
information processing, 49
injuries, 16
injury, xi, 116, 117, 124, 131
insects, 105
integration, 105
interface, x, 80, 95
interference, 29
interphase, 45
inversion, 119, 120
iron, 5
isolation, 62
isotope, 69
isozyme, 72
issues, 96
iteration, 105

L

laboratory tests, 90
landscape, vii, x, 64, 70, 95, 113
languages, 106
lead, x, 34, 76, 128, 133
lead pollution, 133
legislation, 121
legume, 71
lichen, 38
light, 3, 29, 32, 33, 34, 35, 38, 102, 103
lignin, 44
lipid peroxidation, 132
lipid peroxides, 117
lipids, xi, 45, 49, 116, 117
liquid chromatography, 121, 130, 134
local conditions, 109
logging, 67
low temperatures, 3
lying, 20

M

macronutrients, 130
magnitude, 47, 54
majority, 54, 55
man, 119
management, vii, x, 2, 5, 70, 95, 96, 99, 102, 104, 107, 108, 109
mannitol, 47, 48
mantle, 45
mapping, x, 8, 95, 99, 105, 112, 133
marsh, 54, 65
masking, 112
mass, 84
materials, 121
matrix, 62
measurements, x, 8, 72, 95, 99, 103, 105, 107, 109, 111
media, 60, 61
medicine, 92
melting, 46
membranes, 49, 117
metabolism, xi, 61, 116, 129
metabolized, 45
metabolizing, 62
metals, xi, 116
meter, 102, 107
methanol, 60, 62
microcosms, 53, 64, 67, 72
microhabitats, 4, 7, 9
microorganisms, viii, 41, 47, 54, 62, 64, 68
microscopy, 53, 54, 62, 65
military, 102
mixing, 131
models, 80, 97, 105, 109
moisture, 46, 112
molecular biology, 66
molecular weight, 47
morphology, 8, 69

Index

mortality, 2, 29, 30, 31
mycelium, 45, 48, 58, 66, 72, 73
mycorrhiza, 54, 64, 65, 66

N

natural disturbance, 39
next generation, x, 95, 99, 103, 109
nitrite, 72
nitrogen, ix, 42, 44, 49, 59, 61, 66, 70, 72, 117, 120, 121
nodules, 49, 59, 71
nutrient, 44, 46, 66, 116

O

objectivity, x, 76
obstacles, 31
oil, 45, 50, 55, 60, 63, 68, 69
openness, 36
operations, 102
opportunities, 96, 103
ordinary differential equations, x, 96, 107
organic compounds, 48, 69
organic matter, ix, 42, 71
oxidation, xi, 53, 59, 61, 67, 116, 117
oxidative stress, xi, 116, 122, 124, 129
oxygen, 46, 56, 62, 63, 117
ozone, xi, 111, 116, 117, 120, 127, 128, 131

P

parallel, 87, 88
pasture, 51, 70
pathogens, 48
pathways, 61, 68
peat, 65, 72
permeability, 87
permission, iv
phospholipids, 49
phosphorus, 49
photographs, x, 95, 99, 100
photosynthesis, 45, 47, 67
phylogenetic tree, 43, 68

phylum, 51, 65
physical properties, 83
physics, 78
physiology, 66, 77, 132
plankton, 50
polysaccharides, 49
pools, 65
poor performance, 106
population, 16, 18, 37, 54, 55, 56, 58, 60, 63, 64, 65, 69, 109
population size, 109
positive relationship, 30
precipitation, 31, 130
preparation, iv
prescribed burning, 68
preservation, 96
principles, 98, 132
producers, 66
project, 98, 109, 113
prokaryotes, 69
propagation, 109
protection, vii, 1, 5, 97, 106
proteins, xi, 65, 72, 116, 117

Q

quality control, 121
quantification, 70, 111

R

radar, 103
radiation, 126, 132
radicals, xi, 116, 117
rainfall, 4
rainforest, 35
real time, 70
recognition, 104
regeneration, vii, viii, x, 1, 2, 3, 4, 18, 19, 20, 28, 29, 30, 31, 32, 33, 34, 35, 36, 37, 38, 39, 40, 96, 106
regrowth, 35
remote sensing, x, 95, 96, 97, 102, 103, 104, 105, 107, 109, 110, 111, 112

requirements, 3, 33
researchers, 77
resistance, 83, 91, 116
resolution, x, 68, 95, 99, 100, 101, 102, 107, 109
resources, x, 39, 96, 97, 99
respiration, 44, 45, 63, 67, 69
response, 4, 28, 33, 35
rhythm, 77
rice field, 51, 63, 69
rings, 87
rules, ix, 75, 77, 78, 92, 104, 105, 109

S

scarcity, 19
schema, 106
science, 102, 105
scientific publications, 76
scope, 103
sea level, 119
seasonal growth, 73
sediments, 50, 56, 70
seed, 3, 4, 29, 30, 31, 37, 108
sensing, x, 95, 99, 102, 103, 104, 105, 106, 107, 109, 110
sensitivity, 77
sensors, x, 95, 99, 101, 102, 103, 104, 109
shade, 3, 33, 35
shape, 90
shortage, 106
showing, 87, 90
shrubs, 30
silver, ix, 42, 44, 47, 55, 59, 64
simulation, x, 38, 96, 105, 106, 107, 113
skin, 130
society, 96
software, 9, 99, 109, 121
soil pollution, 132
soil type, 107
solubility, 62
solution, 107
sorption, 76, 78, 84, 85, 86
sowing, 77
specialists, 3

species, viii, ix, xi, 2, 3, 4, 5, 28, 29, 30, 32, 33, 35, 37, 39, 42, 44, 45, 46, 47, 48, 53, 56, 58, 59, 63, 64, 69, 79, 89, 100, 108, 109, 110, 111, 112, 113, 116, 119, 134
spindle, 121
sponge, 50
state, x, 77, 80, 95, 107
storage, 97, 98
structure, 2, 7, 15, 16, 18, 37, 38, 39, 47, 66, 67, 70, 71, 72, 103, 112
structuring, 37
subgroups, 50
substrate, 4, 14, 26, 32, 40, 67
substrates, 4, 26, 48, 56, 60, 62, 63
succession, x, 36, 38, 70, 96, 106, 108
sulfate, 70
survival, 16, 29, 36, 38, 107, 108, 109
sustainability, 96
symbiosis, 49, 66, 72, 73
synthesis, 117

T

taxa, 49, 68
techniques, x, 66, 95, 103, 105, 107, 109, 111, 112
technologies, 97, 109
technology, 97, 102, 105
temperature, 4, 64
testing, 38, 104
texture, 112
tides, 77, 92
time series, 107
topology, 97
toxic substances, xi, 115
traditions, 76, 77, 92
training, 104
transcription, 65
transformation, 9, 71
translation, 65
translocation, 66
transparency, x, 76
transverse section, 87
treefall, 36
trial, ix, 76, 82

turnover, 31

U

uniform, 28
universal test machine, 87
urban, 98, 99, 100, 101, 112
urea, 51

V

variations, vii, 77, 78, 79, 80, 81, 82, 83, 84, 88, 90, 92, 94, 131
vegetables, 117, 119, 134

vegetation, vii, viii, x, 1, 2, 3, 5, 7, 8, 12, 14, 18, 21, 28, 32, 34, 35, 36, 37, 44, 95, 100, 102, 103, 105, 107, 108, 110, 112, 119
visualization, x, 96, 98, 105

W

wavelengths, 102
wildlife, x, 96

Y

yeast, 60, 61